D0699578

Brain and Culture

Brain and Culture

Neurobiology, Ideology, and Social Change

Bruce E. Wexler

A Bradford Book
The MIT Press
Cambridge, Massachusetts
London, England

© 2006 Massachusetts Institute of Technology

All rights reserved. No part of this book may be reproduced in any form by any electronic or mechanical means (including photocopying, recording, or information storage and retrieval) without permission in writing from the publisher.

MIT Press books may be purchased at special quantity discounts for business or sales promotional use. For information, please email special_sales@mitpress.mit.edu or write to Special Sales Department, The MIT Press, 55 Hayward Street, Cambridge, MA 02142.

This book was set in Stone Sans and Stone Serif by SNP Best-set Typesetter Ltd., Hong Kong and was printed and bound in the United States of America.

Library of Congress Cataloging-in-Publication Data

Wexler, Bruce E.
Brain and culture : neurobiology, ideology, and social change / by Bruce E. Wexler.
    p. cm.
"A Bradford book."
Includes bibliographical references (p.) and index.
ISBN 0-262-23248-0 (alk. paper)
1. Social change—Psychological aspects. 2. Culture—Psychological aspects. 3. Neurobiology–Social aspects. I. Title.
HM831.W48   2006   612.8—dc22   2005056284

10 9 8 7 6 5 4 3 2 1

In memory of my father, Jack Wexler, whose journey I continue.

Contents

Acknowledgments

I wish to thank all my teachers for making this book possible. Among them, my family has been particularly important as teachers, companions, and supporters: my parents, Ruth and Jack, my wife, Laura, my son, Thomas, and my daughter, Rebecca, my brothers, Richard and Steven, and their wives and children. I offer special thanks to some of my academic teachers who shaped and enabled my thought at critical times and in particularly important ways: Rogers Albritton, Oliver Sacks, Marcel Kinsbourne, and Robert Shulman. I am grateful to Tom Stone, my editor at MIT Press, for seeing the value in the manuscript and then skillfully piloting it through the narrows, and to Ruth Haas, who copy edited with broad knowledge and valuable suggestions.

Brain and Culture

Introduction

Knowledge about the ways our minds and brains work has increased dramatically over the past 100 years. Some of this new knowledge comes from studies of other animals with which we share aspects of behavior. Other information comes from new methods of observation and experimental study of human beings, both as individuals and in groups. Each body of knowledge has its own methods of inquiry and language to collect and describe information, and each represents distinct communities of investigators and theoreticians. Despite this, the different bodies of knowledge are assumed to be fundamentally linked. Neurological and psychological function are two sides of a coin, and different aspects of each are joined in the organic wholeness of the individual. In this book, I explore links among several of the major bodies of new knowledge about human neurological and psychological function. Two things are particularly striking about these links. First, the links among bodies of knowledge that heretofore have remained largely separate are straightforward. Second, important new aspects of human individual and social behavior become apparent through these links.

The first sets of knowledge I consider relate to the period of development between birth and early adulthood. During this period the brain depends upon sensory stimulation to develop physically, and the functional and structural organization of the brain is strongly influenced by the nature of that stimulation. The period of brain maturation and associated environmental influence is much longer in people than in other animals, and the parts of the brain that most distinguish the human brain from those of other primates are the last to mature and subject the longest to shaping by the environment. Stimulation from other members of the species (e.g., parents, siblings, and peers) is a particularly important factor in shaping neuropsychological development. Primate infants seek out maternal stimulation, and numerous studies have demonstrated that such stimulation is essential for the physiological stability and physical growth of the infant.

Research on the neurobiology of social bonding and attachment demonstrated the evolutionary appearance of new mechanisms to support such behaviors in mammals. Research in molecular genetics identified mechanisms through which maternal stimulation of infants creates lasting changes in the structure and configuration of DNA that then influence the level of activity of specific genes throughout the individual's life. Research on human parenting has documented the remarkable degree to which the mother and her infant become an integrated dyadic unit in which the infant develops. L. S. Vygotsky, writing from Russia as a developmental psychologist, and Sigmund Freud in Vienna as well as subsequent American psychoanalysts arrived at virtually identical conclusions about the role of social interaction in creating internal psychological structures. This wonderful expanse of neurobiological, psychological, and social-psychological knowledge all rests on the deep and extended sensitivity of the human brain to shaping by psychosocial and other sensory inputs.

Two important implications emerge when these bodies of work are considered together. The first is the great increase in functional variability among individuals that results from environmental influences on development of the brain. There is an evolutionary advantage for life forms that reproduce sexually because the mixing of genetic material from parents produces variety in their offspring. Thus, different individuals have different characteristics, which increases the likelihood that some members of the group will be able to function and reproduce even when the environment in which the group lives changes. In an analogous manner, the distinctive postnatal shaping of each individual's brain function through interaction with other people, and through his or her own mix of sensory inputs, creates an endless variety of individuals with different functional characteristics. This broadens the range of adaptive and problem-solving capabilities well beyond the variability achieved by sexual reproduction.

The second implication is even more important. In addition to having the longest period during which brain growth is shaped by the environment, human beings alter the environment that shapes their brains to a degree without precedent among animals. These human alterations in the shared social environment include physical structures, laws and other codes of behavior, food and clothes, spoken and written language, and music and other arts. In recent decades, children in Euro-American societies have been raised in almost entirely human-created environments. It is this ability to shape the environment that in turn shapes our brains that has allowed human adaptability and capability to develop at a much faster rate than is possible through alteration of the genetic code itself. This transgenerational shaping of brain function through culture also means that processes that govern the evolution of societies and cultures have a great influence on how our individual brains and minds work.

The second set of knowledge I consider relates to the period of life beginning with young adulthood when the degree of neuroplasticity and the associated environmental shaping of brain function is much reduced. The greater recovery of function after brain injury in children than in adults, and the greater ease with which children learn new languages compared with adults, have long been recognized as signs of a decrease in brain plasticity after childhood. More recent research has shown that the chemical mechanisms of neuronal growth and learning that are so active during childhood are much less evident in adult brains, and learning in adults depends largely on different cellular mechanisms. On the psychological level, established perceptual, attitudinal, and cognitive structures resist change. People selectively perceive and more highly value sensory input that is consistent with their internal values and organizing schemata. People selectively affiliate with like-minded individuals, and forget and discredit views and information inconsistent with their existing beliefs. On the level of subjective experience, people like things more simply because they have seen them more and they more closely match established internal representations. Because of this, individuals generally try to surround themselves with familiar objects and people, and resist intrusions of foreign elements into their environments.

There is, then, a fundamental shift in the relationship between an individual and his or her environment during late adolescence and early adulthood. During the first part of life, the brain and mind are highly plastic, require sensory input to grow and develop, and shape themselves to the major recurring features of their environments. During these years, individuals have little ability to act on or alter the environment, but are easily altered by it. By early adulthood, the mind and brain have elaborately developed structures and a diminished ability to change those structures. The individual is now able to act on and alter the environment, and much of that activity is devoted to making the environment conform to the established structures.

In both periods, however, there is a neurobiological imperative that an individual's internal neuropsychological structures match key features of his or her external environment, a principle of internal-external consonance. This principle is of great importance since people are linked to their environments by obligatory, continuous multimodal sensory stimulation. It is not possible to turn off the sensory receptors. The developing nervous system requires sensory stimulation in order for brain cells to live and develop. Sensory deprivation experiments in adults demonstrate that the developed brain requires sensory input to function properly. These same experiments demonstrate that people seek out even meaningless stimulation to avoid sensory deprivation. Such scientific findings are consistent with the casual observation that after a full day at work, people rarely go home and sit quietly in the dark, but are likely to watch hours of television, listen to music, or seek out social interactions. These obligatory links between the adult individual and the environment have many implications in everyday life. For example, the hormonal state and self-image of sports fans change as a result of the success or failure of the sports team and players with whom they identify. These obligatory links also mean that individuals cannot escape awareness of "foreign" elements in their environments that do not match their established internal neuropsychological structures.

Two implications of this view of the relationship between adults and their environments are of particular importance. The first relates to the change across generations in human society and capabilities. Young adults act on the world to make it conform to their internal structures. Since these internal structures result from each individual's own mix of experience, and from the human-made cultural environment of their developmental years, the internal structures of each generation of young adults differ from those of their parents. When young adults act to change the environment to match their internal structures, they struggle with their parents' generation for control of the public space and, to the extent that they succeed, they alter the rearing environment of their own children. Thus, while their action on an individual level can be seen as conservative in that it aims to alter the environment to match their internal structures rather than altering the internal structures, it is dynamic on the social level because it creates new features of the rearing environment. The continuing nature of intergenerational struggle is also clear from this perspective.

The second implication of the changed relationship between the individual and the environment after young adulthood is in understanding the difficulties individuals face when their environment changes beyond their ability to maintain a fit between internal structure and external reality. One example is bereavement. The work of mourning after the loss of a spouse is painfully difficult and typically takes a full year. It is necessary to restructure a wide range of internal representations of the external world so that they now match an environment of which the deceased is no longer a part. Another example can be found in the difficulties that arise from the meeting of different cultures. A clue to these difficulties comes from the experience of immigrants to a new culture who suddenly find themselves in an environment that does not match internal structures modeled on the rearing environment in their native land. A common response is to recreate a microcopy of their native culture in their homes and friendship circles. Still, like bereavement, it is a prolonged and difficult struggle to reshape internal structures to match the new, general cultural environment. The children in immigrant families are more successful at the internal transformations, often leading to heightened and problematic differences between immigrant parents and their children. Similar issues arise in the contact zones between different cultures and contribute to the interethnic violence that is so common in the world today. Because of the neurobiological importance of the fit between internal structure and external environment, cultures will fight to maintain control over the symbolic environment in which they live and which shapes their children.

The chapters that follow provide the details that support and more fully elucidate the implications of the ideas just outlined. There is little more in the way of argument and few surprises. There are, however, many details. The details are there to make sure that every step in the argument is fully supported. I confess that I like details and the tapestry of integration that they create. I also think readers are entitled to details. Repeated summaries are provided along the way to help the forest remain visible. Still, it is easy to imagine that at times the details may be overwhelming. To understand the argument, it is not necessary to take in all the details, and this introduction, the chapter summaries given here, and the reviews at the beginning and end of the chapters are there to help the reader maintain his or her bearings.

As a neuroscientist and a psychiatrist, I draw on both experimental research and clinical experience as I explore the nature and implications of the rich, lifelong, and biologically mandated interaction between the individual and his or her physical and social environment. My conceptual starting point and foundation, however, are biological, and the first chapter provides a contemporary view of the structural and functional organization of the human brain. The chapter describes general principles of cerebral functional organization and is meant to anchor the book in the particular physical reality of the human central nervous system.

Chapter 2 reviews the research literature on sensory deprivation, beginning with studies demonstrating the dependence of the mammalian nervous system on sensory input to shape its growth and development, and continuing with studies showing the altered function of adult nervous systems in unusual sensory environments. These data are directly germane to the centuries-old philosophical debate [see ref. 1] regarding the extent to which sensory input shapes the structure and content of the mind, and the extent to which inherent qualities of the mind provide categories of perception and thought. While from a biological point of view both seem to be true to at least some extent, the data on sensory-dependent neuronal growth provide clear evidence of the necessity of sensory input for the development of the brain and of the extent to which that input shapes the particulars of human brain structure and function. These data minimize the boundary between the brain and its sensory environment, and establish a view of human beings as inextricably linked to their worlds by nearly incessant multimodal processing of sensory information.

Chapter 3 focuses on the special role of social interaction in development of the brain. This literature comes from diverse starting points, including studies by neurobiologists on the molecular mechanisms through which maternal behavior has lasting effects on the structure and expression of genes in their offspring, Russian neuropsychologists on the role of social interaction in the development of brain functional systems, primatologists on the effects of child–parent and sibling interaction on subsequent behavior, psychoanalysts on the role of similar interactions in the development of internal psychic structures, and cognitive scientists on the development of cognitive function. These studies extend the more general discussion of sensory stimulation in chapter 2, and focus on the special cases in which stimulation comes in the form of another human being or from objects created by human beings. The idea is to demonstrate how interpersonal processes become internalized in the cognitive and psychic structures of an individual, a process with potentially progressive effects across generations.

Chapter 4 reviews evidence that once internal structures are established, they turn the relationship between internal and external around. Instead of the internal structures being shaped by the environment, the individual now acts to preserve established structures in the face of environmental challenges, and finds changes in structure difficult and painful. The chapter reviews evidence that internal neurocognitive structures developed by environmental input then alter the individual's experience of the environment. Individuals seek out stimulation that is consistent with their established internal structures, and ignore, forget, or attempt to actively discredit information that is inconsistent with these structures. Things are experienced as pleasurable because they are familiar, while the loss of the familiar produces stress, unhappiness, and dysfunction. The literature on bereavement illustrates what happens when a major component of an individual's internal and external worlds disappears from their environment, necessitating changes in the internal structures if they are to match the now-changed environment. The experience of immigration is then reviewed as an "experiment" in which individuals are removed from the cultural environment that shaped them and placed in a new cultural milieu. As with bereavement, the subjective discomfort and functional compromise that result from a sudden misfit between internal structure and external environment is striking. Here again, the effort to change established structures to match the new environment is lengthy and painful.

Chapter 5 extends the discussion of immigration to consider the problems inherent in a meeting of different cultures. Since individuals develop internal neurocognitive structures that are consonant with their own culture, the appearance in their environment of individuals from a foreign culture, thinking and acting differently, creates an uncomfortable dissonance between internal and external realities. The first-encounter and travel literatures of the seventeenth to nineteenth centuries provide evidence of nonviolent efforts to reduce this dissonance, efforts that parallel those by individuals confronted by dissonant information within their own culture which are described in chapter 4. The elimination of the Albigensian culture of southern France in the beginning of the thirteenth century by the Catholic Church, the 200 years of war between the Christian crusaders and the Muslim communities in the Middle East, the continuing battles between Muslim and Christian communities in Europe, and ethnic slaughter in Rwanda are discussed as situations, on the other hand, in which the unsettling reality of difference was an important factor contributing to violence.

The epilogue begins with a survey of contemporary conflicts among cultures: the continuing disappearance of small cultures, with devastating impact on millions of human lives; numerous violent conflicts between neighboring cultures; and increasing conflict between the culture of the United States and other cultures it encounters as it spreads throughout the world. It then returns to the interpersonal processes that constitute family life, but with an emphasis on the relationship between these processes and a changing sociocultural milieu. Parents and children are major components of each others' internal neuropsychological structures and external environments. Children, however, are also subject to developmental influences other than their parents, including neighboring and invading foreign cultures. These foreign influences on development ensure that children and parents will differ from each other and from important aspects of the other's internal structures. The pain of Tevye's failed invocation of "tradition" in Sholem Aleichem's story and popular musical "Fiddler on the Roof," as he attempts to prevent his daughter from abandoning customs, can be understood by parents in changing communities around the world. The pain is so great for him that he would rather mourn her loss as if she had died than be confronted by such a prominent feature of his external world in conflict with fundamental components of his internal world.

As culture itself is made into a commodity, and its distribution driven by the combined forces of economics and ideology, the need to understand the intense human response to the interpenetration of cultures is ever more urgent. The epilogue concludes by raising the question of whether the turmoil, excitement, violence, and rapid change associated with the meeting and mixing of cultures in the current epoch of human development will be followed by a homogeneous and static global culture as imagined by George Orwell in *1984* and Ray Bradbury in *Fahrenheit 451*, or by continued change and diversity owing to the exposure of developing brains to an unprecedented variety of influences through electronic information sources.

To make these arguments, this book cuts a straight and narrow swathe through a large expanse of intellectual and scientific territory. All of the areas considered have been visited before, and many have been carefully charted. In the 1960s Clifford Geertz [2] offered the argument that our prehuman evolutionary ancestors had developed cultures, that these cultures participated in the processes of natural selection that led to the phylogenetic development of modern *Homo sapiens*, and that this influence created postnatal human developmental processes that depend upon a cultural milieu. He wrote:

A cultural environment increasingly supplemented the natural environment in the selection process so as to further accelerate the rate of hominid evolution to an unprecedented speed . . . [through] which were forged nearly all those characteristics of man's existence which are most graphically human: his thoroughly encephalated nervous system, his incest-taboo based social structure, and his capacity to create and use symbols . . . [this] suggests that man's nervous system does not merely enable him to acquire culture, it positively demands that he do so if it is going to be functional at all. [3, p. 67].

He goes on to offer a prediction that has since been borne out in the sad reports of children raised in near-total isolation: "A culture-less human being would probably turn out to be not an intrinsically talented though unfulfilled ape, but a wholly mindless and consequently unworkable monstrosity" [3, p. 68].

While I arrive at the same conclusion as Geertz did about the dependence of human development upon a human cultural milieu, I do so through review of a developmental neuroscience literature that arose largely after Geertz wrote. This literature demonstrates the dependence of brain development upon sensory and social stimulation without reference to the origins of this dependence. It elucidates neural and social mechanisms that could underlie Geertz's suggestions, but does not depend upon them.

Lewontin, Rose, and Kamin [4] have articulately marshaled argument and data against an often overly narrow view that human behavior is determined by biological and genetic factors to the exclusion of significant social and environment influence. They write that "humanity cannot be cut adrift from its own biology, but neither is it enchained by it" [4, p. 10]. They conclude that "our task . . . is to point the way toward an integrated understanding of the relationship between the biological and the social" [4, p. 10]. In this book I present an extensive array of neurobiological, psychobiological, and psychological research data that provide a rich picture of the relationship between the biological and the social. These data demonstrate that our biology is social in such a fundamental and thorough manner that to speak of a relation between the two suggests an unwarranted distinction. It is our nature to nurture and be nurtured.

Michael Cole has brought a keen theoretical sophistication and methodological rigor to the study of the influence of culture on psychology, and revitalized the field of cultural psychology in the West [5]. He has discussed the issues arising from awareness of the existence of foreign cultures that are the focus of the second half of this book, beginning with a highly informative discussion of the history of the use of the word "barbarian" to which I refer later. Just as important, he provides a cogent and concise history of psychology that documents the persistence of a "second psychology," begun by the work of Wundt and continued by Durkheim, Levy-Bruhl, Mead, Dewey, and others, which postulated and sought to study the effects of sociocultural structures on individual psychological function. This "second psychology" found particularly powerful proponents in the great Russian psychologists L. S. Vygotsky and A. R. Luria, who themselves thought and worked in the Marxist-Leninist political-cultural milieu that explicitly emphasized the importance of historically evolving social institutions in the thought and function of individuals. In addition to drawing on the work of Vygotsky and Luria, I attempt to integrate their ideas with the developmental neuroscience and interpersonal psychologies of attachment theory and psychoanalysis that developed during the same time periods in England and the United States.

Many others have identified the powerful role cultural differences can play in initiating and maintaining violent conflict between neighboring peoples [e.g., 6–11]. Mary Pratt has coined the term *contact zone* to describe that meeting place between cultures that often turns into a battlefield [10], and Samuel Huntington has suggested that "conflict between civilizations will be the latest phase in the evolution of conflict in the modern world" and that in the current and coming era "the great divisions among humankind and the dominating source of conflict will be cultural" [11, p. 22]. In this book I review data to support the proposition that the importance of culture in shaping the development of the brain, the discomfort that results from disjunctions between established internal structures and a changing cultural environment, and changes in neuroplasticity through the life-span provide a new basis from which to consider these suggestions that differences in culture are an important source of violent conflict. Discussion of the research and writing on bereavement, immigration, first encounters among cultures, and the early European travel literature provide links between the neurodevelopmental literature and culture-conflict theory.

Jared Diamond has written of the demise of societies due to major changes in the physical environment [12]. In some cases the problems were confounded by the way people changed the environment. Many of these human-made changes were based on experiences—and associated internal structures—gained in environments different from the ones in which the people were living at the time of the societal collapse. They represent, at least in part, efforts to make the new environment match internal representations of the old environment. In all instances, the societies were unable to recognize or adapt to the major environmental changes that confronted them. Diamond offers several reasons why the societies he analyzed were unable to prevent calamities that seem obvious in retrospect: inability to recognize problems with which they had no prior experience; denial of information inconsistent with established behaviors and world view; application to new situations of inappropriate analogies based on previous experience; and conflict between actions needed to adapt to environmental change and established social or cultural values. Chapters 4 and 5 of this book discuss these same factors in relation to difficulty adapting to changes in the interpersonal and cultural environments. These examples, and the neurobiological foundation provided in the initial chapters, help explain why these factors can be so powerfully disabling.

Among contemporary controversies in thinking about biology, psychology, and society, the one that my thoughts most clearly address is about the relative roles of nature and nurture in shaping human development. In this book I review and integrate a wide range of information about the role of nurture or, more accurately, the external environment. I suggest that the phylogenetic emergence of human beings rests, to a significant degree, on selection for an extended period of postnatal plasticity in the fine-grained shaping of the structural and functional organization of the human brain. This is a nature and nurture argument. The extended period of postnatal neuroplasticity is an aspect of human nature that allows and requires environmental input for normal development. Moreover, since human beings provide an important part of the environmental input, one person's nature is another person's nurture. There are also aspects of human nature that constrain and define specific aspects of human development and function, and important research is continuing to better characterize and understand those processes. For example, the studies I present in chapter 2 that provide evidence of the effect of rearing environment on aspects of intelligence also indicate that heredity has a large effect on intelligence test scores. And when I discuss differences among cultures in chapter 5, I do not mean to suggest that there are not also commonalties across cultures that might be associated with inborn aspects of human nature.

The goal of this book, however, is to present a full view of the powerful ways in which the environment in general, and the human and human-made environment in particular, influence human development, and then to discuss the implications of these influences for understanding individual and societal behavior. I think such a review can contribute in valuable ways to the ongoing discussion about nature and nurture; to my knowledge, such a review and integration of these factors and processes does not currently exist. The fact that I do not also review the ways in which inborn factors influence human development should not be taken to mean that I think such factors do not exist or are not important.

When entering as many realms of observation and inquiry as I do in developing the theses offered in this book, I do not treat each as fully as I might if it were the focus of the enterprise. This limitation is most pronounced when I consider the sociopolitical consequences of differences among cultures, and suggest, as have others [e.g., 6–11], that these differences have contributed in important ways to violence and warfare. If an analysis of the causes of violence and war was itself the focus, other important factors would need to be considered.

On the other hand, the route followed affords its own opportunities. This book draws together a wide range of data that support and explicate the role of postnatal neurodevelopment in the remarkable evolutionary success of the human species. These data describe neurobiological mechanisms capable of supporting the social and cultural influences on individual growth and function described by Geertz, Vygotsky, Cole, and others, and provide a more contemporary biological basis for these earlier developed theories. In doing so, they provide a counterbalance and context for the excitement produced by new work characterizing the human genome and the processes of gene expression and regulation. Based on the findings of research on the brain and psychological development, I have been able to articulate the principle of internal-external or neuroenviron-mental consonance. On the basis of this principle, the brain shapes itself to the external physical, social, and cultural milieu in which it develops during the early years of life. Later in life, on the basis of this same principle, the individual seeks out an external environment that matches the already established internal structures. When the available external environment does not match the internal structures, the individual acts to alter the environment to make it match the internal structures. Finally, while the primary reason for considering such a wide range of information is to buttress the overall argument by support from multiple quarters, in doing so, links are established between distinct areas of inquiry and bodies of knowledge. I hope that these links will enrich the reader's understanding of information from all sectors.

1 Background: Some Basic Facts about the Human Brain

This chapter briefly summarizes some facts about the human brain that are particularly relevant for understanding the ways in which the environment affects brain structure and function. I begin with a description of the ways in which neurons, the cellular building blocks of the brain, form the multicell ensembles and multiensemble systems that constitute the basic functional units of the brain. I then consider mechanisms of learning and memory through which the brain is altered to create lasting representations of its environment. Next, I describe three aspects of the brain that appear relatively late in phylogeny and are of particular importance in understanding the human brain and human culture: the proliferation of cell number, especially in the frontal and parietal lobes, which distinguishes human beings from even our nearest primate neighbors; the extent of postnatal brain development that again distinguishes human beings from other primates; and the emergence of brain structures that support the familial and other social behaviors that distinguish mammals from other animals. These latter structures, and their associated familial and social behaviors, are also the basis of emotion, and the final section of the chapter briefly considers the neural basis and neurodevelopmental significance of emotion.

The Neuron

Neurons are individual nerve cells. There are 100 billion in the human brain, and they are functionally linked with one another by a combined chemical and electrical communication system. Each neuron has thousands of receptors or chemical "docking stations" on its outer surface or membrane. Chemicals released from one neuron attach to the receptors on neighboring neurons. When the chemicals, known as neurotransmitters or neuromodulators, attach to a neuron, they initiate a series of chemical reactions that alter the electrical state of the neuron's outer membrane. Some of these reactions increase the voltage differential across the cell membrane, while other reactions decrease the differential. When this voltage differential reaches a critical value, an electrical signal is transmitted down the length of the neuron. This electrical wave then causes the release of neurotransmitter or modulator molecules that affect the electrical state of other neurons.

In this way, neurons serve as functional building blocks of information-processing modules. When the combination of excitatory and inhibitory inputs to an individual neuron reaches the right mix, the neuron fires its electrical system and sends chemical signals to its neighbors. The complexity of the networks of functionally interconnected neurons is almost beyond comprehension. On average, each of the 100 billion neurons receives direct input from a thousand other cells and may receive hundreds of signals from other neurons within a millisecond. Moreover, some scientists think that there are functional processes in addition to neurotransmitter release and electrical transmission that influence the state and firing of groups of neurons, adding more dimensions of complexity.

Figure 1.1 is an illustration of neurons and the dendritic and axonal projections that interconnect them. Two important points are relevant for this book. First, processes such as thinking, remembering, and feeling arise from the integrated action of many neurons and are not properties of individual neurons. Second, the specific patterns of all the intricate connections among neurons that constitute these functional systems are

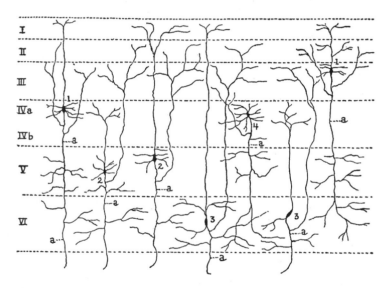

Figure 1.1

Schematic drawing of nerve cell bodies (marked by numbers) and the long axons (marked by "a") and shorter dendritic branches that interconnect them. The interconnecting branches extend across layers of the cerebral cortex that are defined by populations of different types of nerve cells. The cells are actually much more densely packed than indicated in the drawing, and all the dendritic branches make contact with other cells. (Source: RC Truex and MB Carpenter, *Human Neuroanatomy*, 6th ed., Lippincott, Williams and Wilkins 1969.)

determined by sensory stimulation and other aspects of environmentally induced neuronal activity. In the details of this neuronal "wiring," no two individuals are the same.

Multineuronal Functional Units

It was previously thought possible that the functional basis of mental processes such as thinking, feeling, and remembering would somehow be found in individual neurons. This idea, parodied in the idea of a "grandmother" cell that contained the memory of one's grandmother, including the cookies she made and the hugs she gave, has given way to the idea that functions like thought, feeling, and action depend on the integrated action of many neurons and are not properties of individual cells. Many scientific studies have supported this view. For example, studies recording activity in single brain cells during sensory stimulation demonstrate that many different cells each respond to many different stimuli [1,2]. Individual cells may respond to repeated presentations of the same stimulus with enough regularity to establish a relationship between cell and stimulus, but there is too much variability in response from one presentation to the next to make an individual cell a reliable indicator of a particular stimulus [1–3]. The required consistency is found in the behavior of large groups of cells rather than the behavior of single neurons. As a consequence, large numbers of cells respond to even simple sensory stimuli, and large numbers of cells change their response characteristics after learning even simple associations between stimuli.

In one study, conditioning by association of a picture with a reward led to changes in the electrical response to the picture in thirty of thirty-two cells from which activity was recorded

[1]. Another study [4] used radioactive tracers in the sugar consumed by the brain to measure localized learning-related increases in neuronal activity in cats. First, the investigators surgically separated the two halves of the brain (cerebral hemispheres) so that one hemisphere could serve as a control for the other. Next, the cats were trained to recognize a geometric shape that marked the door behind which food was to be found. Using colored figures and contact lenses with appropriate filters, the learned stimulus was then presented to one cerebral hemisphere while a novel stimulus was presented to the other. A comparison of sugar consumption in the hemispheres indicated that over five million cells distributed throughout the brain showed learning-related responses to a single, simple geometric form. Of course, many aspects of the environment become linked to one another through experience, and numerous additional brain functional imaging studies have also demonstrated widespread activation responses to simple sensory stimuli [see ref. 5] for particularly nicely controlled demonstrations). Similarly, large populations of neurons participate in individual behavioral acts, and the populations for different behaviors overlap [6].

Multicellular functional units such as those suggested here are consistent with an architecture that connects each neuron with hundreds or even thousands of others. Indeed, each of the 100 billion neurons in the human brain is thought to be connected to all the others through pathways with no more than six intermediary cells. This is not to suggest, however, that all neurons, or all functional ensembles of neurons, are the same or functionally equivalent. Simply on a physical level there are different types of neurons, and the mix and organization of the different cell types varies from region to region throughout the brain. Moreover, structurally similar cells have different patterns of interconnection and different response characteristics. It is these differences among neurons that allow myriad functions and representations to emerge in their organization as functional or representational units, just as letters of an alphabet are organized to form the different words of a language. The same structural units can have very different functions when they are arranged differently (e.g., tea, eat, ate). Moreover, existing functional units can be incorporated into larger units, giving up their original function entirely as they contribute to the new function (e.g., team, meat, mate). It is these types of functional units, in which functional properties emerge from the organization and integration of differentiated components, that are particularly important in creating the characteristics that distinguish the human brain from the brains of other animals.

The simplest neurofunctional unit to exhibit properties not seen in single neurons is the two-neuron reflex arc. Indeed, Sherrington predicted the existence of distinct neurons that interact on the basis of the discrepancies he observed between the reflexive motor response to electrical stimulation of a sensory nerve and the electrical properties of a single nerve. (The reflex motor response is slower than the conductance time of a single nerve; it shows a response after the cessation of a stimulus; and it is more subject to fatigue, refractory periods, and the effects of drugs.) When tens, hundreds, or even thousands of neurons are linked, more complex and powerful functional properties are created. Moreover, an endless variety of such modules can be created, each with somewhat different characteristics. These modules can themselves be combined to form systems that constitute capabilities for such processes as speaking, reading, writing, thinking, and remembering.

Based on work with hundreds of Russian soldiers who had suffered many varieties of brain injury, Luria collected data that are consistent with this type of functional organization in the human brain. He observed that

a disturbance of a particular complex function does not in fact arise in association [only] with a narrowly circumscribed lesion of one part of the cortex, but is observed as a rule . . . with lesions of several different parts of the brain. Disorders of writing . . . may appear in lesions of temporal, post central, premotor and occipito-parietal regions, . . . [of] objective movements with . . . parieto-occipito lesions and frontal lesions, . . . [of] reading with occipital, temporal, and frontal lesions of the left hemisphere . . . and so on. [7, p. 12].

Thus he concluded that each of those regions was part of a system necessary for the particular behavior. Moreover, "a lesion of a narrowly circumscribed area of the cortex practically never leads to the loss of any single isolated mental function, but always to the disturbance of a large group of mental processes" [7, p. 13]. This led him to the conclusion that each cortical region contributes to multiple different behaviors and multiple cortical regions contribute to each individual behavior.

Changes in Brain Functional Organization during the Course of Life

Luria also concluded from his work that the same function could be carried out by a different collection of neuronal modules in different individuals, or in the same individual at different times. Direct evidence of this can be seen in the fact that similar brain injuries in similar locations have different immediate functional consequences in children than in adults. Apparently,

through the course of development, neuronal modules themselves are reconstituted, and/or the same modules come to play
different functional roles. Luria found further evidence of plasticity in the relationship between brain structure and function
in patients who recovered functions that had been lost as a
result of the irreversible destruction of some part of their brains.
The initial loss of function demonstrated that up to that point
in time, the destroyed area of the brain had been crucial for the
function. Recovery of the function without recovery of the
destroyed area of the brain demonstrated that the function was
now based on a different collection of neuronal modules.

Functional recovery in adult patients with brain injuries is
slow and arduous, and, sadly, usually limited in extent. This is
part of the evidence that structure–function relationships can
change much faster and to a much greater extent in infant
animals and prepubescent children than in adults. For example,
researchers have surgically rerouted visual input in newborn
ferrets to the part of the brain that usually processes auditory
information [8]. In this case, the animals developed fully functioning visual processing systems using what are naturally the
"hearing" modules of their brains. And in young children who
have had the entire left side of their brain surgically removed
for treatment of otherwise untreatable seizures, all functions
become localized in the remaining right hemisphere. Even the
language function, which would otherwise have involved cyto-
architectonically specialized regions within the left hemisphere,
becomes based entirely upon right hemisphere structures. In
contrast, adults who suffer damage to language-specialized areas
in the left hemisphere are often left with major and permanent
impairments in language function. There has been research suggesting that it may be possible to enhance functional recovery

by making more effective use of the neuroplastic potential of the adult brain, or even to enhance this potential pharmacologically. Even so, neuroplasticity in the developing brain during childhood is very much greater than it is in the adult brain.

Microscopic Demonstration of Experience-Based Modification of Interneuronal Connections

In the 1940s Hebb proposed that modification of connections among neurons was a basic mechanism of learning, remembering, and other environmentally induced changes in brain organization [9]. Changes in the strength of interneuronal connections as postulated by Hebb have since been microscopically demonstrated and quantified in a series of studies in snails (*Aplysia californica*) [10]. The scientist who conducted these studies, Eric Kandel, was awarded the Nobel Prize for this work.

The primary extension into the environment of the snail's body is called its gill, and includes the structure with which it feeds. If the gill is touched lightly, it reflexively withdraws from the touch. The neuronal system responsible for the reflexive withdrawal consists of sensory neurons that connect directly with motor neurons and with small "interneurons" that in turn contact the motor neurons. If this harmless tactile stimulation is repeated, the reflex response progressively diminishes. Electrical recordings from the motor neurons during this process indicate that the excitatory electrical changes produced in the motor neurons by the sensory neurons become smaller. Chemical analyses show that there is a decrease in the number of neurotransmitter molecules released from the sensory neurons. These changes occur at several locations in the reflex circuit, and in this sense the memory that the light tactile stimulus is harmless is distributed through the circuit. After only four experimental sessions, each consisting of merely ten harmless tactile stimuli to the gill, the number of sensory neurons with physiologically detectable connections to motor neurons decreased from 90% of those sampled to only 10%. With further presentation of the mild tactile stimuli, there was an actual physical pruning of the branches of the sensory neurons that contact the motor neurons, with a one-third reduction in the number of physical contact points between sensory and motor neurons, and a fourfold reduction in the proportion of contact points with physiologically active zones (figure 1.2).

Just the opposite sequence of events is observed if the gill withdrawal reflex is activated by potentially harmful electric shocks applied to the snail's tail. When the shocks are repeated, neurotransmitter release from the sensory neurons is enhanced. This in turn leads to release of nerve growth factors and a more

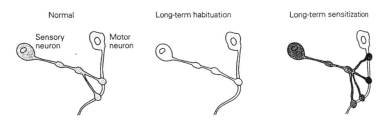

Figure 1.2
Schematic representation of reductions in the number of connections between nerve cells as a result of repeated presentation of mild, harmless sensory stimuli (long-term habituation), and increases in connections as a result of repeated presentation of potentially harmful stimuli (long-term sensitization). (Source: E. Kandel et al., *Principles of Neural Science*, 4th ed., McGraw-Hill, 2000.)

than twofold increase in the physical contact points between sensory and motor neurons (figure 1.2). Both studies demonstrate the remarkable extent of environmentally induced reshaping of neuronal interconnections with even simple stimulation over a short time in a simple nervous system.

Activation of a Response System by Partial Inputs

Even before Hebb, Kuhlenbeck wrote that "an engram [a memory trace] represented a patterned association of neurons, caused by a permanent modification of synapses [the contact points between neurons], and spreading throughout an extensive region" [11, p. 101] Kuhlenbeck's idea was that the effect of synaptic alteration is to create a particular pattern of associations among neurons and that it is the pattern that has representational value. He went on to suggest that the discharge of such a pattern of neurons could be triggered by a fractional input. Since then, computer scientists have created models of "neural networks" that are able to learn by altering the connectivity among network components. Consistent with experimental studies of information processing, and with Kuhlenbeck's prediction, these networks: (1) recognize many different examples of a particular category of objects after exposure to only a few members of that category; (2) recognize a prototype of a particular category of objects more accurately than particular examples of that category, even when the prototype itself has not been previously presented; and (3) recognize objects following the presentation of only part of the object.

In these model networks, and it is thought in the brain itself, information, knowledge, and skills are represented in multimodular functional and representational systems that develop through lasting modification of connections among units produced by interaction with the external world. This aspect of brain development is not predicated on an inborn or preexisting correspondence between the brain and the external world, but through this development the brain is altered to bring about such a correspondence. Consistent with the results of research on perception in both human beings and other animals, representational patterns in these model systems are activated by stimuli that bear only partial or incomplete resemblance to the stimuli from which the representational patterns are derived. This provides efficiency in recognition of novel presentations of familiar objects, but at the same time leads to an increasing influence of internal structure on perception in such a way as to reinforce the internal structure.

There are two key points here. The first is that brain functions such as thinking, feeling, and remembering result from the integrated action of many brain cells that through networks of interconnection form functional systems. Second, sensory stimulation and its associated neuronal activity create the networks of connections among neurons, and thus shape the functional systems. Much more scientific data illustrating and supporting this second key assumption will be presented in chapter 2.

Differences between the Brain of Human Beings and the Brains of Other Animals

For heuristic purposes, the human brain can be divided into components that derive roughly from three major stages in evolution [12]. The first is a collection of cell groups and tracts that are found in the brains of all reptiles, birds, and mammals. Referred to by MacLean as the reptilian complex, these are located at the base of the human brain. The second is the cortex and its associated collection of nuclei called the limbic system that immediately surrounds the core "reptilian" brain and is present in all mammals, but not in reptiles. The third is the neocortex, which surrounds the limbic system and which has shown progressive phylogenetic development throughout the mammalian line, reaching an unprecedented size in the human brain. The cortex and neocortex are often referred to collectively as the cortical brain regions.

The primary difference between the human brain and the brains of phylogenetically more primitive mammals is an increase in the size of the neocortex. The human neocortex is more than 200 times larger (relative to body size) than that of phylogenetically early mammals. (The reptilian complex and limbic system have shown relatively less growth.) Within the neocortex, growth in the human has been most marked in the frontal and parietal lobes, with the size of some sensory receptive areas reduced relative to the overall neocortex compared with other mammals [13]. Increases in cortical volume come in part from an increase in the branches of neurons that interconnect them. The overall increase in the human cortex comes primarily, however, from a marked increase in surface area rather than cortical thickness, and here the primary process is an increase in the number of neurons. The average surface area of the human cortex is 3.5 times larger than that of the orangutan, an ape approximately the same size as human beings. The inferior parietal lobule is nearly 9 times larger and the frontal lobes more than 5 times larger [14]. With regard to the relative proportions of different brain regions, the prefrontal cortex is about 24% of the cerebral mantle in humans compared with 14% in the great apes [15], and the relative enlargement of the parietal lobe is associated with a relative reduction in the occipital lobe to half the size expected in a primate with a brain as large as the human's [16]. The increase in the size of the neocortex is also associated with changes in the laminar organization of the cortex.

These observations, then, provide additional support for the view that the special functional properties of the human brain come from an increase in the number of basic components already present in the brains of lower mammals, and from increases in connectivity and changes in organization among these components. It is of particular interest that these increases are greatest in the frontal and parietal lobes of the brain, regions that in human beings continue to mature and develop into the third decade of life. In contrast, corresponding structures in the brains of chimpanzees and other higher primates reach comparable levels of maturity by the second or third year of life. Thus, high levels of plasticity in the relationship between structure and function persist for years in the structures that most distinguish the human brain from those of other primates. This creates an unprecedented opportunity for environmental shaping of uniquely human aspects of brain function.

Familial and other interpersonal interactions are particularly important aspects of the shaping environment, as will be discussed in chapter 3. The biological foundation for these interactions rests on the limbic system. The evolutionary appearance of the limbic system is associated with the evolutionary appearance of parenting and other familial behaviors. Surgical lesions in the rat limbic system decrease nest building and other maternal behaviors and result in a marked increase in mortality of offspring, despite having no effect on the mother's survival or on most aspects of her behavior. Similarly, the distress cry of infants separated from their mother or nest, which is typical of most infant mammals, is reduced or eliminated in squirrel monkeys that have lesions of the limbic system. Moreover, while repeated studies of electrical stimulation throughout the neocortex of monkeys have consistently failed to generate any vocal behavior, stimulation at multiple sites in the limbic system produces distinct, naturally occurring vocalizations. Noting reports by human patients who had subjective experiences of emotions during seizures in the limbic cortex, MacLean further linked limbic structures with emotion [12].

Emotion

Limbic-based familial and social behaviors precede the development of language, both in evolution and in the course of human infancy and childhood. These behaviors depend upon innate, nonverbal facial and vocal displays of emotion. Both facial and vocal displays of the basic emotions are highly similar and easily recognizable across cultures [17]; people as distinct as illiterate New Guinea tribespeople who have never seen television or other photographic media, and contemporary young Japanese and Americans all have highly similar, mutually recognizable, distinct expressions for happiness, anger, disgust, sadness, fear, and surprise. Clear similarities are recognizable between human and other mammalian facial and vocal displays of emotion [e.g., 18]. Moreover, children who are born without sight show the same displays of emotion on their faces as do children with normal vision, and children born deaf use the same vocal expressions of affect as do hearing children [19]. As expected from their association with limbic structures, spontaneous emotional displays are present in neurologic patients with facial motor paralysis secondary to vascular lesions in the motor areas of the neocortex, but are disrupted in patients with subcortical limbic lesions that do not affect deliberate movements of the face.

Emotions themselves are of particular importance in the present discussion because of their role in social interaction and their effects on brain and body systems. Emotions transform our relationship with the world [20], and we are more likely to think and act in specific ways associated with specific emotional states [21]. Brain imaging studies have found changes in activity in many regions throughout the brain following the presentation of emotion-evoking stimuli (reviewed in ref. 22). Recent work in my laboratory found that when people watched videotapes of an actress smiling and talking about happy things, they felt happy themselves and the functional links among regions in their brains were very different than when they watched videotapes of an actress crying and talking about sad things and felt sad themselves. The essence of emotion is this conjunction of the display behaviors and the modification of internal brain and body processes. Emotion is an interindividual process that alters the momentary functional organization of the brains of the interactants, configuring and activating certain multiunit functional systems and dismantling and deactivating others.

Contagion is at the heart of emotion. Fear spreads rapidly through a herd of antelope, activates specific behavioral predispositions, and enables all the individual animals to behave in parallel or act in common. When one person sees a photograph of another person smiling, electrical activity increases in the viewer's zygomatic muscles, which turn the corners of the mouth up into a smile. Looking at a photograph of a person frowning leads to an increase in electrical activity of the viewer's corrugator muscles, which furrow the brow in a frown [23]. This automatic mimicry of facial displays is one way that we know what those around us are feeling. It also serves to create the perceived emotional state in the viewer. When people deliberately create facial expressions of fear, anger, surprise, disgust, sadness, and happiness, changes occur in heart rate and skin temperature that match those that occur when they remember and relive each emotion [24]. When subjects exaggerate their facial responses to electric shocks, there is an associated increase in skin conductance and subjective responses to the stimuli [25]. Subjects who contract their eyebrows (thus creating a frown) while watching cartoons rate the cartoons as more angry and less happy than do subjects who lift the corners of their mouths (thus creating a smile) [26]. These studies demonstrate that deliberately created emotion-related facial expressions alter subjective and physiological responses to sensory stimuli.

Spontaneous facial expressions that occur during the course of social interaction have the same effects. Multiple brain areas are more active when subjects are looking at pictures of people's faces than when they are looking at other objects (reviewed in ref. 27), and some are more active still when people deliberately imitate rather than simply observe the facial expressions [28]. Two of these, the orbitofrontal cortices and the amygdala, are of particular interest. Both are part of the limbic system; both show greater responses to negative than positive emotion-evoking facial displays within one-fifth of a second of stimulus presentation; and damage to either results in impaired ability to recognize facial expressions [27,29]. The amygdala even shows greater activation to facial displays of fear than to happiness when the facial stimuli are presented subliminally and processed without conscious awareness [30]. Adults comfortably and regularly create exaggerated facial displays of emotion when interacting with infants and young children, and this is one of the ways in which they alter brain activity in their children. It has been estimated that between the ages of 3 and 6 months, infants are exposed to over 30,000 exaggerated facial displays of emotions [31]!

Summary

The functional units of the human brain consist of modules that each include thousands of nerve cells connected to one another via networks of branches. Functions such as perception, recall, and thought arise from integrated systems of modules distributed throughout the brain. Thus injuries to a particular location in the brain affect multiple functions, and each function is affected by injuries at multiple locations. The specific connections among nerve cells are determined by sensory stimulation and other environmentally induced neuronal activity, and differ uniquely from person to person. The human brain is most distinguished from the brains of other primates by the number of brain cells and the patterns of their interconnections.

The two regions of the brain that differ most between humans and other primates are the frontal and parietal lobes. These structures are 5 to 9 times larger in humans and remain highly susceptible to organizational shaping by the environment for 5 to 10 times longer than in primates. The brains of human beings and other mammals also differ from the brains of phylogenetically older animal species in having a number of structures together called the limbic system. These structures provide the basis for familial and social behaviors and for preverbal visual and vocal displays of emotion. Emotion is an interindividual process that alters the moment-to-moment functional organization and activation patterns of the brain in the individuals who are interacting.

I Transgenerational Shaping of Human Brain Function

2 Effects of Sensory Deprivation and Sensory Enrichment on Brain Structure and Function

The relationship between the individual and the environment is so extensive that it almost overstates the distinction between the two to speak of a relationship at all. The body is in a constant process of gas, fluid, and nutrient exchange with the environment, and the defining feature of each body organ is its role in these processes. The brain and its sensory processes are no exception. Sensory input is always a physical interaction with the environment. Borrowing a trick from plants, retinal cells in the eye capture and transform photic energy using molecules that change shape when they are exposed to light. These changes in molecular shape initiate a series of changes in the cell membrane that in turn initiate an electrical discharge. Sensory receptors in the ear and skin convert the mechanical effects of movement into electrical impulses. Specialized cells in the mouth and nose have receptor molecules on their surfaces that combine with molecules from the environment to initiate electrical impulses to the brain.

Individuals often have an exaggerated sense of the independence of their thought processes from environmental input. This is due in part to the nature of memory, which allows individuals to carry in themselves the effects of environmental input to places distant in time and space where the origins of the effects drop from awareness. It is also because we are not capable of detecting, tracking, and summing the subtle and numerous environmental influences on our development and thought, any more than we can count and track the different molecules in the air we breathe or the food we eat. As psychotherapists and their patients know, it is only with great effort and in the unusual circumstances of the psychotherapy process that some of the experiential antecedents of our present selves can be identified, and then they remain available to conscious introspection only briefly. Thus, as we develop into unique individuals as a result of both our unique cumulative interactions with the environment and our unique hereditary characteristics, our uniqueness seems a property of us.

This chapter and the next aim to establish the degree to which the brain is usually dependent upon input for development and function. For this purpose it is sobering to consider that from one perspective, the brain is surpassed by the digestive system in regard to autonomy in its interactions with the environment. The stomach efficiently takes apart a wide range of inputs, reducing them to components that contain no reference to their initial organization, and then delivers them to the body for reconstruction according to the individual's needs. In contrast, the brain recreates in itself a representation of environmental input which, especially during formative years, conforms highly to the complexities of that input.

Experimental study of the role of sensory input in brain development and function has proceeded along several avenues. A series of studies of animals, conducted primarily during the

1960s and 1970s, examined the effects of sensory deprivation, beginning immediately after birth and extending for weeks or months, on the development of brain structure, functional organization, and functional performance. One line of subsequent studies looked at the effects of unusually stimulating environments on these same aspects of development. Another has sought to determine the extent of functional reorganization possible in the adult brain as a result of sustained changes in sensory input. Studies of the effects of sensory deprivation on development of the human brain have been limited to evaluations of individuals with congenital sensory deficits or early-life, disease-related sensory losses. The effects of an enriched environment on the development of the human brain are evident in the effects of rearing environment seen in intelligence tests. Finally, a relatively large experimental literature demonstrates the dependence of the adult human brain on sensory stimulation for maintenance of established functions. This chapter reviews each of these areas in turn.

Animal Studies

Effects of Sensory Deprivation on Brain Structure

The brain's dependence upon environmental input begins with the need for sensory stimulation to maintain structural integrity. Information-processing structures along the entire information input pathway, from peripheral sensory receptors to cortical processing centers, atrophy when they are deprived of sensory input. The number of ganglion cells in the retina, which carry excitation from the photoreceptor cells of the eye to the first relay station in the brain, is decreased to 10% of normal in dark-reared chimpanzees, and cats and rats have smaller than normal ganglion cells after dark rearing [1]. Electron microscope examination of the eyes of chicks reared for 4 weeks with opaque coverings of one eye revealed morphological abnormalities in the rod and cone photoreceptor cells themselves in the deprived eye [2].

Similar sensory dependence is seen in the first relay station for visual information in the brainstem, the lateral geniculate body, and in the area of the cortex that receives input directly from the lateral geniculate body. Both the number and size of cells are reduced by as much as 30–40% in the lateral geniculate of cats and monkeys deprived of visual input during the initial weeks of life [e.g., 3–9]. The effects continue along the information input pathway to the visual cortex, where cell number, cell size, and the density of connections among cells are decreased and the organization of cells is altered [e.g., 10–16]. Studies of olfactory deprivation have yielded a similar picture. Occlusion of one nostril in rat pups leads to a decrease in volume of the region of the brain to which neurons from that nostril project (the olfactory bulb), with increased cell death, decreased cell number, and decreased connections among remaining cells [e.g., 17–19].

The effects of sensory deprivation on structural integrity can be prevented by injecting a nerve growth factor into the cerebrospinal fluid within the brain during the period of deprivation [20–24]. This naturally occurring substance is most likely "a target-derived neurotrophic factor the production or uptake of which [is] under the control of afferent electrical activity" [22, p. 25]. Thus, sensory-induced activity along a particular pathway promotes growth and connections among the neurons along that pathway.

Effects of Sensory Deprivation on Functional Organization of the Brain

The effects of sensory deprivation on the development of the brain's functional organization are more profound even than the effects on brain structure. In general, these effects result from changes in the balance of activity in deprived and nonde-prived sensory pathways. As a result, for example, limiting input to both eyes has less effect on functional organization of the visual system than does limiting input to only one eye. Neurons at each stage of processing compete for connections with the neurons at each subsequent stage, with the neurons that fire more often gaining territory.

These effects were investigated systematically in Hubel and Wiesel's Nobel Prize-winning studies of kittens and monkeys [8]. They recorded electrical activity from individual cells in the area of the cortex that receives visual information from the lateral geniculate relay station; i.e., cells that are just one neuron away from the eye itself. By recording from cells when visual stimuli were presented first to one eye and then the other, they determined the extent to which each brain cell responded to input from each eye. By recording from hundreds of individual brain cells in each animal, they determined that in animals raised under normal conditions, most cells respond to inputs from both eyes (approximately 85% in the kitten, 65% in the monkey). Many of these cells responded somewhat more frequently to input from one eye or the other, with such preferences divided evenly between the eyes. Similarly, of the monocularly responsive cells, half responded exclusively to the right eye and half to the left. However, when an eye was sutured closed shortly after birth and then reopened approximately 10 weeks later, a very different picture emerged. Eighty-five percent or more of the cells responded preferentially to the previously nondeprived eye, and few if any cells responded exclusively to the previously deprived eye. Responses to stimulation of the previously deprived eye were slow to start, decreased in amplitude, and easily fatigued when they were present at all.

When visual input to the deprived eye is restored, the altered pattern of cortical cell sensitivities persists despite the fact that both eyes are now receiving unobstructed visual input. As long as neurons from the previously nondeprived eye remain active, they are able to maintain their abnormally acquired hegemony. If, however, the previously nondeprived eye is occluded while the animal is still young enough, the abnormal response pattern can be normalized or reversed in favor of the previously deprived eye. Apparently the deprivation-induced functional reorganization is based on reversible changes in neuronal projections and connections rather than on cell death or the elimination of fiber tracks [8].

Experimental alteration of stimulation through other sensory modalities has similar effects. For example, the removal of whiskers from infant rats or mice leads to expanded areas of cortical response to stimulation of the remaining whiskers [25, 26]. When the clipped whisker is allowed to regrow, stimulation activates a smaller than normal cortical area [27]. When sensory input from an area of skin is eliminated by injury to the nerve arising from that area, reorganization of the cortical somatosensory map results. Cortical cells that were originally responsive to input from the damaged nerve now respond to stimulation on new, usually adjacent areas of the skin [e.g., 28]. The potentially extensive anatomical range of such functional reorganization is suggested by both electrophysiological recordings and measures of metabolic activity that demonstrated responses to whisker stimulation in cells within the visual cortex of adult rats who had an eye removed at birth [29].

Critical Importance of Form or Content in Stimulation

When one eye of a kitten is covered with plastic that allows diffuse light but not discrete forms to reach the retina, the effects on brain functional organization are similar to the effects of suturing the eye closed. This is so despite the fact that the plastic covering only reduces light by 50% compared with the 10–50 fold decrease following suture of the eyelid. Thus elimination of form perception rather than light itself seems to be the critical factor. Similarly, experimentally altering the nature of the form (i.e., content) of the visual stimuli changes the functional organization of the visual cortex even when the stimuli are viewed normally by both eyes. For example, some cells in the visual cortex respond selectively to moving objects, with each cell having maximum sensitivity to movement in a particular direction. Kittens raised in a strobe light that prevents appreciation of movement have decreased numbers of motion-sensitive cells [30, 31]. Presumably cells that would have been "specialized" for detecting movement became selectively responsive to some other aspect of visual information instead. Kittens raised in the dark except for exposure to stripes moving from left to right have a marked increase in the proportion of cells selectively responsive to left-right rather than right-left movement [32].

Other cells respond selectively to lines (i.e., an object's edges), with each of these cells having maximum sensitivity to lines of a particular orientation. Kittens exposed to vertical black-and-white stripes for a few hours each day, but otherwise reared in darkness, have cortical cells with preferences for vertical line orientation's, but none with preferences for other orientations [33]. Kittens raised wearing goggles that allowed them to see only vertical lines in one eye and horizontal lines in the other have fewer than the normal number of cells that respond to oblique lines. Moreover, the cells responsive to vertical lines are active only with stimulation of the eye that had been exposed to vertical lines, and the cells responsive to horizontal lines are active only with stimulation of the eye that had been exposed to horizontal lines [34].

Cells with the same response selectivities tend to be located together and interconnected in clusters that extend through several layers of cerebral cortex; these are called ocular dominance or iso-orientation columns. In the kitten, such clusters normally emerge in crude form from an initially unclustered circuitry during the second postnatal week, or shortly after natural eyelid opening at postnatal day 7. By 4 weeks the clusters have typically reached the degree of refinement seen in adult cats. This refinement, however, is prevented by deprivation of binocular vision. In deprived animals, numerous cells between clusters show the same responses as cells within the clusters, apparently because they have maintained collateral axonal connections to these cells that are eliminated in the normal course of development [35].

These studies were all carried out, and awarded the Nobel Prize, because the scientific community thinks the findings apply to our brains as well as to the brains of kittens and monkeys. The same types of studies cannot be done in people, but observations described later in this chapter of children with certain medical conditions that limit sensory input support the assumption that the results of the animal studies apply also to the development of the human brain. These studies provide evidence, then, that our brains (and minds) develop concrete perceptual structures, capabilities, and sensitivities based on prominent features of the environment in which we are reared, and then are more able and more likely to see those features in the sensory mix of new environments we encounter. Or conversely, we have limited ability to see even prominent features of a new environment if those features were absent from our rearing environment.

Effects of Sensory Deprivation during Infancy on Brain Function

The structural and organizational effects of sensory deprivation during infancy have functional consequences that are easily demonstrated in laboratory experiments. Most directly, there is an increase in perceptual thresholds. After olfactory deprivation, higher concentrations of odors are necessary for detection, and the ability to recognize individual odors is decreased [36]. Rats deprived of auditory input are at a disadvantage compared with rats deprived of visual input when an auditory stimulus is used to start a competition for food, while the reverse is true with a visual starting signal [37]. Visual acuity is decreased on tests requiring discrimination of stationary objects [e.g., 38, 39] and after monocular deprivation is decreased severalfold more on tests requiring detection of horizontal displacement of a section of slowly moving parallel lines [Vernier acuity; ref. 39). The latter discriminations are normally much finer than the former, and are thought to depend upon the pattern of divergent and convergent projections from cells in the lateral geniculate to cells in the visual cortex. Dysfunction in this aspect of visual acuity presumably results from the abnormal pattern of functional interconnections among cells that follows from periods of monocular deprivation.

Of more interest, the effects of early bilateral sensory deprivation are also evident on more complex information-processing tasks that are not themselves dependent on stimulus detection thresholds. For example, rats deprived of auditory input for the first 60 days of life are still able to discriminate pitch normally, but are less able than normally reared rats to learn sound patterns or tonal sequences [40]. Dark-reared rats are impaired on tasks requiring judgment of duration, perception of depth and discrimination of form. In the study of durational judgment, the rats received food pellets for pushing one lever following a short auditory or visual stimulus, and for pushing another lever following a longer stimulus. Dark-reared rats did as well as controls with auditory stimuli, but significantly worse with visual stimuli [41]. Depth perception was assessed by placing rats in a box, the floor of which consisted of a center board flanked by panels of glass. Under each glass panel was a back-illuminated surface that could be raised and lowered. When the surface under one panel was lower than the surface under the other, animals with good depth perception more often walked off the center board onto the side with the shallower rather than the deeper "visual cliff." Normally reared rats reliably chose the shallower side when the difference between sides was as small as 9 cm. Rats raised in the dark for their first 90 days made consistent distinctions when the difference between sides was 45 or 27 cm, but not when it was only 18 or 9 cm. Rats dark reared for 150 days did not show distinctions, even with a 45-cm differential [42]. Discrimination of form has been assessed by training animals to respond differentially to two forms; for example, an "X" as opposed to an "N," or a circle as opposed to a triangle.

Dark-reared animals are slow to learn such distinctions [e.g., 43, 44], and show no evidence of functional recovery despite extended periods of normal visual experience after the period of deprivation [e.g., 42]. Moreover, after having learned to distinguish "N" from "X," for example, previously visually deprived animals are less able than controls to recognize these same forms when they are rotated, made smaller, or presented white on black instead of black on white [42].

Electrophysiological recordings following bilateral visual and auditory deprivation parallel the findings of increased sensory thresholds. Kittens raised in darkness show no electrical responses following presentation of contour-varying visual stimuli [45], and rats raised in sound-shielded rooms required louder auditory stimuli to produce cortical electrical responses [46]. More complex functional abnormalities have also been revealed by electrophysiological recording. For example, by inserting stimulating electrodes in the olfactory input pathway, and recording electrodes in the olfactory bulb, it is possible to demonstrate that the response in the bulb to the second of two paired stimulations is lower than the response to the first. This "prepulse inhibition" is increased in rats deprived of olfactory input during rearing. Citing the fact that prepulse inhibition is known to also be regulated by neurons that project onto the bulb from other cortical centers (i.e., centrifugal inputs), the authors suggest that "reorganization of these and other centrifugal inputs following deprivation-induced loss of normal target cells could result in a change in the excitability of the system" [47, p. 190].

Equally important, visual deprivation also decreases the ability of sustained stimulation of a particular neural circuit to enhance the response of that circuit to subsequent brief stimuli. Such long-term potentiation is thought to play an important role in certain types of learning and memory [48–50]. It has been proposed that the initial sustained stimulation leads to structural and functional enhancement of synaptic connections, which in turn are the basis for enhanced response to subsequent brief stimuli [48–50]. This augmentation of response has been demonstrated by implanting both stimulating and recording electrodes in the brains of animals and comparing responses at the recording electrode with brief stimulation through the stimulating electrode before and after sustained stimulation through this electrode. The effect can also be demonstrated in slices of brain tissue kept viable in a bath of artificial cerebrospinal fluid and an atmosphere of moist, oxygenated air. Slices from the brains of the light-deprived rats showed significantly less long-term potentiation in the cells of the visual cortex than did slices from the brains of rats raised as usual.

Thus, we see more evidence that sensory stimulation is necessary if the brain is to develop normally; the link to the environment is essential rather than elective. Moreover, the effects of sensory stimulation extend past the initial steps of perception and affect brain systems that are related to the evaluation and retention of sensory input.

Effects of an Enriched Rearing Environment on Brain Structure and Function

Another way to see the effects of the environment on the developing brain is by comparing animals reared in more versus less stimulating environments. In some studies the more stimulating or enriched environment is a cage with playthings and perhaps other animals, while the control condition is an empty, dimly lit cage. In others it is rearing in the wild, while the impoverished environment is the standard laboratory cage. In each study the two conditions differ with regard to the overall amount of ambient stimulation, but it is only relative to each other that one is enriched and the other not. These studies all differ from the sensory deprivation experiments in several respects. First, no single sense is completely (or even almost completely) deprived of input in the low-stimulation condition. Instead, there is a general impoverishment of several types of sensory input. Second, the differences between the conditions that are compared in an enrichment study are less marked than the differences between conditions in sensory deprivation experiments. Third, the conditions being compared tend to differ more in the degree of complex, multimodality stimulation and sensorimotor experiences than they do in raw sensory stimulation. Most important, the enriched-environment experiments compare conditions that are more similar to situations that might actually exist in the natural world.

With regard to brain structure, the results of enriched-environment experiments have been qualitatively similar to those of the sensory deprivation experiments. Animals raised in relatively impoverished environments have smaller brains, with the reduction greatest in the cerebral cortex and unrelated to differences in body weight [e.g., 51, 52]. There is less protein synthesis in multiple regions of the cortex [53], decreased area of synaptic contact among neurons [54], and decreased numbers of the axonal and dendritic branches that functionally connect neurons [55].

The functional consequences of enriched rearing environments have been dramatically, albeit indirectly, demonstrated in studies of the effects of experimentally induced brain injuries. Animals raised in enriched environments show fewer functional deficits and recover more quickly following such injuries, suggesting that their increased axonal and dendritic interconnections provide a functional reserve, and/or the basis for a rapid functional reorganization [e.g., 56–58]. There are also numerous direct experimental demonstrations of the functional advantages conferred by an enriched learning environment. Compared with the sensory deprivation literature, these studies demonstrate effects on more complex neurocognitive functions. Studies in rats, cats, and monkeys have found that animals raised in enriched environments are better at learning the relative positions of different objects [59], learning the correct route through a variety of mazes [60–64], learning to discriminate among visual stimuli [65, 66], and learning to avoid noxious stimuli [67]. Such effects are evident in adult animals in whom the exposure to an enriched environment was limited to the period of active growth and development immediately after birth [e.g., 67–69].

Effects of Rearing Environment on Frontal Lobe Functions
Of particular interest, studies in both cats and monkeys have found that animals raised in enriched environments perform much better on tests of frontal lobe function than animals raised in less stimulating environments [69, 70]. Several studies used delayed-response tests in which animals must match an initial visual or auditory stimulus with a second stimulus presented a few seconds after the first. All cats raised in a normal environment, for example, performed well with 1-minute delays between stimuli, and half did well with 6-minute delays. Seventy-five percent of cage-reared cats, in contrast, could not manage even a 15-second delay. Moreover, the deficits were similar with auditory and visual stimuli, indicating the sensory independence of the functional deficit [70]. Such comparisons

require that the first stimulus be kept in a cognitive computational workspace (i.e., working memory), where it is then compared with the second stimulus [71–73]. The ability to make such comparisons is a fundamental aspect of many important cognitive operations, and in humans the amount of material that can be held in working memory is highly correlated with scores on intelligence tests [e.g., 74]. Moreover, it is dependent upon the function of frontal lobe structures. Lesions to the frontal lobes, but not to other brain areas, impair performance on delayed-response tasks [75], and cells have been identified in the frontal lobes that increase their activity specifically during the delay period between the two stimuli [76–78].

Further evidence of the effect of rearing conditions on frontal lobe function is seen in the difference between monkeys reared in more and less enriched environments on a problem requiring planning, strategy, and the mental manipulation of information [79]. People with frontal lobe injuries or tumors are impaired on these functions, and brain imaging studies have shown selective activation of regions within the frontal lobes on similar tasks [e.g., 80]. It appears then that sensory stimulation, and perhaps associated motor activity, can affect the development of the functional capability of frontal lobe systems, the area of the brain that is farthest from direct sensory input and one of the two areas that most distinguish the human brain from the brains of other animals.

Effects of Sensory Deprivation and Environmental Enrichment in Adult Animals

The studies reviewed until now have examined the effects of sensory stimulation on animals during the first weeks or months of life, a period during which the brain changes more rapidly than it does in adulthood, and during which the impact of sensory stimulation on brain structure and function is most substantial. For example, the dramatic changes in functional organization noted with monocular deprivation only result from deprivation during this early critical period, and can only be normalized by reverse occlusion during this time [24]. The end of this period is usually marked by the achievement of sexual maturity. After this point in development, the brain remains dependent upon sensory stimulation for normal function, shows a variety of compensatory responses to the withdrawal of stimulation, and maintains a more limited capacity for structural and functional reorganization in the face of unusual sensory input.

After either surgical or pharmacological disruption of sensory input in an adult animal, the cortical neurons deprived of sensory input increase their spontaneous activity [81]; low-level stimulation of the otherwise deprived areas leads to greater than normal responses; and proliferation of receptor sites is evident on the deprived neurons [82]. It is as if the neurons deprived of sensory input make physiological changes to amplify any available signal, and perhaps also increase the release of trophic factors to strengthen the failing or failed input connections. We see here evidence of a "withdrawal" effect similar to that seen in drug-dependent individuals when they first stop using drugs. The brain is dependent upon sensory stimulation and shows withdrawal effects when it is deprived of that input. We also see evidence of processes that serve to maintain established neuronal structures by maintaining input from established sources.

If the selective sensory deprivation persists, however, there is a functional reorganization of the cortex around and including the deafferented area, even in the brains of adult animals. For example, a localized area of visual cortex can be deprived of excitatory input by making bilateral focal retinal lesions in homologous regions of both eyes. Within minutes, cortical cells that had previously responded to retinal cells within the outer rim of the lesioned area expand their receptive fields and respond to stimulation of retinal cells just outside the lesion area. Cortical cells that previously responded to retinal cells in the center of the lesioned area initially fail to respond to any visual stimulation. However, after a few months these cortical cells become responsive to stimulation of the retina near the lesion site. Limited or no changes in the lateral geniculate body, which relays information from the retina to the cortex, imply that in this case the originally deprived cells now receive stimulation via intracortical connections [83, 84]. Other studies show similar changes in subcortical relay stations [85], demonstrating that under conditions of severe sensory deprivation the adult brain reorganizes at both the subcortical and cortical levels so that cells deprived of their initial sensory input pathways become responsive to other active afferent pathways. Changes of this nature have been documented in many studies of auditory as well as somatosensory and visual cortices (reviewed in ref. 86). They can also be produced by extreme accentuation of a specific aspect of normal sensory experience [87, 88], appear in the motor cortex when nerves to specific muscles are cut [89–91] or individual muscles are activated over and over [92], and can lead to reorganization of brain areas farther down information-processing pathways than the initial sensory receptive areas [93].

Activity-Dependent Changes Can Account for Large Differences among Individuals

In several studies of adult monkeys, the researchers first mapped out in each monkey the area of the brain in which neurons were responsive to stimulation of each digit [e.g., 87]. They found that these maps of brain functional organization varied as much as 75% from monkey to monkey prior to the experimental manipulation. These differences among individuals, however, were smaller than the changes induced by the experimental manipulation in each individual. Based on this, the authors suggest that "use-driven alteration of cortical representations accounts for most of the great normal variability in the details of cortical representations recorded in adult monkeys" [87, p. 100]. In other words, there are large differences among individuals in the area of the brain that receives input from each finger, and these differences appear to be the result of differences in prior sensorimotor experiences.

Different Cellular Mechanisms Underlie Changes in Infant and Adult Brains

Studies of the molecular mechanisms of activity-dependent changes in brain structure demonstrate differences between these processes in developing and in adult animals. Throughout life, sensory-stimulated activation of cortical cells is mediated in part by the neurotransmitter glutamate. Glutamate is released by axonal terminals of the sensory relay neurons from the brainstem and activates cortical neurons by attaching to N-methyl-D-aspartate (NMDA) receptors on the cortical neurons. All, or nearly all, effects of sensory activity on the developing brain are prevented when the NMDA receptors are blocked [94–96]. This is true even with only a partial blockade of the receptor that still allows the cortical cells to be activated by sensory input [96]. Researchers think that in addition to carrying electrically encoded information to the cortical cells, the NMDA receptors directly activate neural growth mechanisms. What is particularly important is that the number of NMDA binding sites in adult animals (cats) is only half that in developing immature animals [96]. Moreover, in 3-week-old kittens, 50% of the response to visual input in layers IV, V, and VI of the cortex is mediated by NMDA receptors, but this figure drops to 10 or 15% in adult cats [96]. Thus this central mechanism of neural plasticity in the developing brain is markedly reduced in the adult brain.

More recent molecular genetic studies show, similarly, that genes activated by exposure to an enriched environment also show decreased expression with advancing age [97], providing further evidence of an age-related decline in the processes that underlie activity-related neural plasticity. Some researchers go further and assert that "the molecular mechanism underlying adult activity-dependent structural plasticity differs [in kind] from the molecular mechanisms underlying plastic changes in the developing brain" [98, p. 608]. This conclusion is based on the observation that when NMDA receptors are eliminated from a portion of the brain in adult mice by genetic manipulation, environmental enrichment still leads to increased neuronal growth in that region. Further work needs to be done, however, to show that this non-NMDA-mediated plasticity is absent in developing mice.

Summary of Studies in Animals

In summary, numerous well-designed experimental studies in animals unequivocally demonstrate the dependence of the mammalian nervous system upon sensory stimulation for growth and development. This dependence begins with the peripheral sensory receptors themselves and extends through the initial sensory relay stations to association and information-processing centers throughout the brain. The survival of individual cells, the number of dendritic branches and synaptic connections among cells, the structural organization of cell groups, the functional response characteristics of individual cells, and the competence of neural functional systems all depend profoundly on the extent and nature of environmentally induced activation.

These experimental results are important for the overall argument of this book in three ways. First, they establish that dependence of the mammalian brain upon sensory stimulation is obligatory. The brain is bound to the environment for survival and function and thus is inescapably subject to environmental influence. Second, environmental stimulation shapes the structural and functional organization of the brain; it is not simply that a predetermined organization requires sensory stimulation to be realized. Moreover, a measure of the importance of environment for organization of the brain is provided by studies in adult monkeys [87]. Careful cell-by-cell mapping of the somatosensory cortex before the experiment revealed marked differences in brain functional organization from one monkey to the next. However, experimental manipulation of sensory stimulation produced changes in individual monkeys that were greater than the differences that existed among monkeys before the experimental manipulation. This demonstrates, as those investigators concluded, that differences in rearing environments can readily account for differences in brain organization among adults. Third, once brain organization evolves, and the individual reaches sexual maturity, existing structures tend to be enduring and resistant to change. This is due in part to changes in brain chemistry that reduce neuroplasticity. It is also because as long as input pathways and neuronal ensembles remain active, the existing organization is stabilized. While activity-related functional reorganization is possible in adult mammals, it is much slower, much more limited, and achieved with much greater physiological effort.

Human Studies

There is nothing to suggest that the human brain would lack the sensitivity to shaping by sensory stimulation that has so consistently been demonstrated in experimental studies of other animals. On the contrary, given that the time from birth to sexual maturity is much longer in humans than in other animals, as are the times before adult levels of sensory, motor, and cognitive function are achieved, one might expect the human brain to be particularly sensitive to the effects of sensory input and environment-induced behavior patterns on development. Moreover, the human brain differs most from those of other animals in the amount of cerebral cortex, and in the animal studies it is the cortex that was most affected of all brain regions by sensory stimulation and enriched environment.

Systematic experimental investigations of the effects of selective sensory deprivation during brain development are not possible in humans. Instead, scientists have relied on "experiments of nature"; partial and temporary sensory deprivations as a result of disease and/or the treatment of disease. These have been primarily in the visual system and include periods of monocular deprivation during childhood. The results in humans are strikingly similar to those in other animals. Additional information is provided by observations of children who have had one cerebral hemisphere removed. Although neither sensory input nor motor activity has been experimentally controlled in these children, the massive functional reorganization of their brains is thought to be activity dependent, and certainly attests to the organizational plasticity of the human brain during development. Recently, functional brain imaging techniques have made it possible to compare brain functional organization in groups of people with very different sensorimotor experiences. These studies have also yielded data similar to those of the animal studies.

Systematic study of the effects of enriched rearing environments on brain development are also not possible in human subjects. Here, though, scientists have taken advantage of "experiments of society," and evaluated the effects of different family rearing environments on tests of intelligence. Again the results are consistent with studies in other animals.

One area of study has been more extensive in humans than in other animals: the effects of sensory deprivation on established aspects of adult brain function. These studies establish that the adult human brain requires ongoing sensory stimulation to maintain ongoing function.

Effects of Sensory Deprivation in Children

Two congenital visual problems are associated with partial and, because of treatment, usually temporary visual sensory deprivation. The first are congenital cataracts, or opacities of the lens of the eye, which severely limit both light and form from reaching the retina. After removal of the cataract, the physical and functional characteristics of the eye can be determined. As in young animals deprived of visual input, the axial length of the eye itself is increased [99]. Also as in animal studies, the threshold for light detection is elevated in the formerly deprived eye [100]. Electrical responses to visual stimuli, measured on the surface of the skull overlying the visual cortex, are decreased in amplitude, increased in latency, and absent with more subtle stimuli [101], and depth perception is seriously impaired [e. g.,102, 103]. Part-time occlusion of the noncataractous eye is a standard part of treatment after surgical removal of the cataract, in order to promote use-dependent enhancement of function in the visual pathways originating in the formerly deprived eye. Acuity and sensitivity do improve over time after removal of the cataract, demonstrating stimulus-dependent enhancement of function in the formerly deprived visual pathway. However, depth perception remains seriously compromised, as in animals after monocular deprivation, presumably because of the failure to develop the normal population of binocularly innervated cells in the visual cortex.

There have been a number of cases reported over the past 200 years in which bilateral congenital cataracts, or bilateral cataracts that developed as a result of early childhood illness, have been removed 20, 30, or 40 years later in adulthood. Because of curiosity about the visual ability of such individuals, and because recovery of function is much slower in these adults than in children, some of these patients have been more extensively studied and described than the cases treated in childhood [104–106]. Several observations are of particular interest. First, for the first days after removal of the cataracts, the patients are unable to make sense out of the visual world, even if the mechanical operation of the eye allows clear vision. This is so in mature, intelligent, literate, competent adults who have known and functioned effectively in a world without sight. The answer to the question posed by William Molyneux to his fellow seventeenth-century philosopher, John Locke, is clear: A blind person who can easily identify objects by touch cannot then recognize those objects when first able to see them. Information gained by some other senses is not sufficient to organize this new visual input. Such organization requires visual experience itself. Second, use of the eyes is not spontaneous, at least initially, and takes instruction, encouragement, and special effort. Apparently, through disuse, the visual system has lost the intrinsic level of activity—stimulus searching and automatic processing—that characterizes even the newborn infant.

Third, after individuals become able to identify details or aspects of an object, they often have difficulty putting them together. As Oliver Sacks said of a patient he examined: "He would pick up details incessantly—an angle, an edge, a color, a movement—but would not be able to synthesize them" [106, p. 123]. These details—angles, edges, color, and motion—are the component perceptual processes of the visual system. A series of studies in monkeys [107] has established that these components are normally synthesized in later stages of secondary visual processing. The examples of prolonged visual deprivation would seem to indicate then that these later stages, and these secondary association areas one or two synapses beyond the primary visual cortex, also depend on sensory input to develop and/or maintain function. Fourth, visual working memory is severely impaired. Even after weeks of vision, for example, individuals cannot keep in mind a string of letters, but instead complain of forgetting the initial letters as they view subsequent ones. These same individuals had excellent tactile and auditory working memories. This deficit in visual working memory in the presence of intact tactile and auditory working memories indicates that disuse of a particular sensory modality causes the failure of development and/or atrophy of neural systems that use information from that modality in cognitive operations. These systems have been shown to normally involve frontal lobe regions (see earlier discussion), probably acting in concert with the primary and secondary sensory areas. Although not conclusive in this regard, the deficits in visual working memory related to prolonged visual deprivation are more evidence that the development of frontal lobe functions is dependent upon sensory stimulation.

Fifth, these patients have great difficulty learning to recognize things visually, even though their general capacity for learning in other sensory modalities is perfectly normal. Sacks's patient, for example, repeatedly confused his dog and cat, which were similar in size and coloring, despite many deliberate attempts to study and master the differences in their appearance. Similarly, a patient studied by Gregory [105] still could not recognize faces or facial expressions a year after surgical restoration of the mechanical aspects of vision.

This limitation in visual learning is evidence of the loss of central visual competence through disuse. Although the nature of the neural loss is not known, the previously cited animal studies suggest several possible mechanisms. For one, the neural resources necessary for learning, and normally available to visual input, may have established exclusive connections with other sensory modalities. Alternatively, or in addition, the cellular mechanisms of long-term potentiation, shown in animal studies to be important in learning and to be decreased in the visual cortex of dark-reared animals, may be permanently depressed after years of light deprivation. In any case, here again is evidence of the effects of sensory deprivation on aspects of brain function other than simple sensory registration. Finally, patients whose vision is restored after lifetimes of blindness complain particularly of difficulty in recognizing even familiar objects seen from a different perspective, of a different size, or seen only partially. Recognition of this sort depends upon the transformational processes that are essential to all sensorimotor functions and probably to many aspects of thought as well. A deficiency in this aspect of visual information processing, while transformational processes are well developed in auditory and somatosensory processing and in motor function, emphasizes again both the extent to which the brain is functionally organized by sensory modality and the extent to which the development of the brain's ability to act upon as well as register sensory input depends upon sensory stimulation.

The second childhood visual disorder related to sensory deprivation is strabismus. In strabismus the two eyes do not focus together. There are two forms. In alternating strabismus, the child focuses (and uses) one eye on some occasions and the other eye on other occasions. Both eyes are used and both maintain visual acuity. However, since the two are not used simultaneously, the normal population of binocularly responsive cortical cells does not develop, and here again depth perception is severely impaired. In other cases of strabismus, one eye is used regularly and the other rarely or never. It is not known how often, or to what degree, the eye not used is less visually acute at birth than the eye that is used. It is clear, however, that by the time these children undergo a formal assessment of visual acuity, the unused eye is much less acute than the used eye. Moreover, if the condition persists untreated, vision in the unused eye deteriorates further and the eye may eventually become blind. The loss of acuity in this eye is evident even when the child wears glasses that provide a sharply focused retinal image, suggesting that it is related to a dysfunction in the neural input pathway or visual cortex rather than in the mechanics of the eye. As in the case of a monocular cataract, occlusion of the preferred eye leads to forced use and increased visual acuity in the formerly ignored eye. However, the degree of improvement is greater if treatment is begun at an earlier age [e.g., 108]. In addition, here again deficits in depth perception persist even when vision becomes comparable in the two eyes and the ability to use both eyes together is achieved through exercise, growth, and/or surgery.

Although these observations are limited to only visual sensory deprivation, the phenomena are robust and have been repeatedly confirmed. They suggest four conclusions. First, in humans as in other animals, monocular deprivation leads to alterations in the eye itself and to decreased function of visual input pathways, including the brainstem relay station and/or the visual cortex. Second, in humans as in other animals, forcing an underfunctioning visual pathway to process visual stimuli leads to functional enhancement of the pathway. Third, in humans as in other animals, the absence of simultaneous stimulation of the two eyes during development leads to a loss of depth perception. Fourth, correction of the sensory input dysfunction leads to greater restoration of function when it is done at an earlier age. Together these observations make a strong empirical case for similarity between humans and other mammals in the role of sensory input on the physical and functional development of the brain.

Effects of Unusual Sensorimotor Experience on the Sensorimotor Cortex

Animal studies demonstrated that changes in sensorimotor behavior caused changes in the organization of the sensorimotor cortex. Brain functional imaging studies in humans show similar effects. For example, one study found that the volume of sensorimotor cortex associated with movement of the index finger was greater in blind Braille readers than in non-Braille-reading controls, presumably in response to the use of that finger for hours every day [109]. Another study used magneto-encephalography to estimate the amount and location of neuronal activity in the sensorimotor cortex when light pressure was applied in an alternating fashion to the first and fifth fingers of one hand and then the other [110]. Two groups of right-handed subjects were compared; one was a group of string instrument players and the other was a group of nonmusicians. All of the musicians used their left hands to finger the strings of their instruments, a task that requires the fingers to make many rapid movements based on tactile cues of pressure and position. On average, the musicians had played their instruments for nearly 12 years and practiced approximately 10 hours per week. The musicians proved to have more intense and more extensive activity in the sensorimotor cortex during stimulation of their left hands than did the nonmusician controls. Moreover, the intensity of response to left-hand stimulation was greater in the musicians who began playing at or before 12 years of age than it was in those who began playing between ages 14 and 19. The groups did not differ in brain response to right-hand stimulation. These and other similar studies provide strong evidence that in humans, as in other animals, extensive sensorimotor experience of a particular kind can alter the brains functional organization, just as does the absence of an ordinary component of sensory experience. Again, these effects are greater when the sensorimotor experience occurs earlier in life when neuroplasticity is greater.

Effects of Rearing Environment on Human Intellectual Function

Many studies in many different countries have consistently demonstrated that the rearing environment significantly affects the development of higher brain functions in humans (see ref. 111 for a full review of this work). These studies are analogous to the studies of enriched environment on the developing brain in animals, although the human studies are much more extensive. Like the animal studies, the human studies provide evidence of the effects of environment on aspects of brain function that go well beyond the early stages of sensory information processing. A variety of different tests of cognitive function have been employed, ranging from the multidimensional Wechsler Intelligence Tests for adults and children to the Raven's Progressive Matrices Test of visual-spatial ability. On the Wechsler scales and several of the others, scores are expressed in relation to the scores of a large comparison group, with 100 being equal to the average score, approximately two-thirds of people being between 84 and 116, and only one out of 50 being above 132. Scores on multiple tests within a test battery like the Wechsler are often averaged to produce a single performance score, and this score then related to the individual's age to produce an intelligence quotient (IQ) score. One set of studies has looked at the effects of birth order and family size on intelligence tests. Another set has examined the effects of adoption on test scores. A third series of studies has documented a steady rise in test scores over the past 100 years, which is thought to be associated with changes in rearing environment.

Firstborn children have higher IQs than their younger siblings on average, and secondborn children generally have higher IQs than siblings that follow [e.g., 112]. One exception is only children, who have lower IQ scores than secondborns from families of two or three children, presumably because the absence of any siblings removes a source of enrichment. As family size increases beyond two, however, there is a progressive decline in IQ for all children in the family. For example, one study of 800,000 high school juniors in the United States found an average score of 105.3 in children from two-child families compared with 100.0 in children from five-child families [113]. A study of 400,000 19-year-olds in the Netherlands found a performance decrease of nearly 6% in the Raven's test in firstborns from a family of five children compared with firstborns from a family of two children, and a decrease of nearly 11% in firstborns from a sibship of six to eight [112]. Both birth-order and family-size effects are thought to reflect the quality of parental attention devoted to the children, especially during the first years of life. Indeed, several studies have documented differences in the way mothers interact with their firstborn children, compared with later-born children. Mothers talk to their firstborn infants more often, use more complex sentences when doing so, look at and smile at their firstborns more often, sustain interactions with their firstborns far longer, play with and give toys and pacifiers to their firstborn more often, and respond more quickly to the overtures of their firstborns [114–117].

The decline in IQ with an increase in family size from two to three children is not seen when the children are spaced at least 3 years apart [111,118], nor is it seen in even larger families in the highest socioeconomic quartile [119]. Both of these observations support the notion that the family size effect is due to environmental factors, although the possibility remains that more intelligent parents have fewer children, plan more time between each child's birth, and earn more money. If so, children from smaller families, or in upper-income large families, may have higher test scores on a genetic basis and not because of a more enriched rearing environment. Other data discussed later, however, cannot be explained on a genetic basis. Nor can the birth-order effect be explained by genetics since comparisons are made among children with the same parents. One might wonder, though, whether the birth-order effect results from the fact that younger mothers, and first-time mothers, are generally in better physical health and therefore provide a better intrauterine environment.

Here, however, the data are clearly in the opposite direction; children born of older mothers have higher IQs. For example, in a study of 49,000 British children, firstborns of mothers 25 years old or older had IQs 8 points higher than firstborns of teenage mothers; secondborn children of mothers 25 years old or older had IQs 6 points higher than secondborns of mothers between 20 and 24 years old; and second- and thirdborns of mothers over 40 years old had average IQs more than 5 points higher than second- and thirdborns of mothers between 25 and 29, and 8 points higher than those of mothers between 20 and 24 years old [120]. Another study of 1,500 siblings in two-child Dutch families found higher test scores in children of older mothers regardless of birth order [121]. This finding makes it very unlikely that birth-order effects within families are due to mothers being younger when they have their first child than they are when they have subsequent children; when birth-order effects are corrected for maternal age, they are even more robust.

The positive effect of rearing environment independent of genetic contributions has been repeatedly demonstrated in studies of adopted children in which IQ has been measured in both the children and their biological mothers and siblings. A study of 63 children adopted during the 1930s in Iowa found that on average the adopted children scored 30 points higher in IQ than their biological mothers [122]. A subsequent larger study of 300 adopted children found the IQ increase limited to children whose biological mothers had relatively low IQs, but then the effect was still robust (a 13-point increase for children whose mothers had IQs of 95 or less; ref. 123). Moreover, a third study found that adopted children scored 16 points higher than their biological siblings who remained with their biological parents [124]. The IQ enhancement in adopted children must result from some feature of the adoptive environment. At the time of these studies, the number of babies available for adoption was far short of the number of families wishing to adopt. Potential adoptive homes were carefully screened, and parents who succeeded in getting children for adoption had considered and sustained commitments to becoming parents. Presumably these processes and features translated into enhanced rearing environments.

Most remarkably, average national scores on a variety of different intelligence tests have risen steadily and substantially ever since the tests were first developed and applied nearly a century ago. This phenomenon was first noted in the 1980s by James Flynn when he was studying the test scores of American military recruits [125]. Flynn noticed that recruits who had average test scores (i.e., 100) compared with their contemporaries were above average when they were compared with recruits who had taken the same test a generation before. He then reviewed the IQ testing literature and identified many studies in which large groups of subjects had been given two different intelligence tests—sometimes older and newer versions of the same test and sometimes two different tests. In each case, the large comparison groups used to establish the average score (i.e., 100) for one of the tests had been tested years before the comparison group for the other test. Invariably, subjects who scored 100 (average) in reference to the newer comparison group scored above 100 in comparison with the older group. For example, more than 1,500 children were given both the Wechsler IQ test (comparison group from 1947) and the Stanford-Binet IQ test (form L, comparison group from 1931–1933). Their average score on the Wechsler was 103.0, while on the Stanford-Binet it was 107.7, implying a gain of 0.33 IQ point per year from 1931–1933 to 1947 in the general population from which the two comparison groups were drawn. Similarly, more than 200 children were given both the 1947 and 1972 versions of the Wechsler IQ test. They scored 8 points higher when they were compared with the 1947 reference population than they did when they were compared with the 1972 reference population, indicating a gain of 0.32 point per year in the general population from 1947 to 1972. By collating results from numerous other similar comparisons, Flynn has determined that the population mean changed at approximately this same rate throughout the twentieth century, with a cumulative change of 24 points between 1918 and 1989 in the United States [125].

Similar changes have subsequently been identified in all other countries where available data have permitted an assessment (e.g., England, Scotland, France, Norway, Sweden, Belgium, the

Netherlands, Denmark, Germany, Japan, Israel [111,125,126]). Prior to appreciation of the increase in test scores in the general population over this period, studies demonstrating lower IQ scores in older than in younger adults had attributed the difference to decline in performance with aging. Now it is appreciated that many of these differences are due to year of birth. Rather than losing capability, the older individuals, because they were born in an earlier period, never achieved the same level of capability as did the younger group with which they were compared. Direct evidence of this has been supplied by a longitudinal study of intelligence in different birth cohorts. The differences again evident between younger and older cohorts are greater than the decline with age seen in each cohort [127,128].

The causes of this progressive increase in IQ over the past 80 or 100 years are not well understood. It is possible that it is in part due to improved health during childhood, related to such factors as better nutrition, improved obstetrical and neonatal care, and the successful prevention of childhood diseases through inoculation, but the magnitude of these effects are generally thought to be great enough to account for only a fraction of the IQ changes [129]. A panel of specialists in intelligence research created by the American Psychological Society concludes that "perhaps the most plausible [explanation] is based on the striking cultural differences between successive generations. Daily life and occupational experience both seem more 'complex.' . . . The population is increasingly urbanized; television exposes us to more information and more perspectives on more topics than ever before; children stay in school longer; almost everyone seems to be encountering new forms of experience" [129, p. 30]. This view is consistent with the substantial difference (6 points) in IQ between rural and urban populations in the United States noted in studies during the 1930s and 1940s [130,131], and a subsequent decrease in the difference (to 2 points) as increased travel and mass communication decreased rural isolation, rural schools improved, and farm technology became more sophisticated [132,133].

It is also consistent with evidence demonstrating increases in IQ related to schooling. For example, when children of virtually the same age go through school a year apart because some have birthdays just before the cutoff for beginning school and others just after, those who have had a year more of schooling have higher IQ scores [134]. When public schools were closed for several years in part of the state of Virginia in an effort to avoid school integration, depriving most black children in the area of any schooling, the IQ scores of the children dropped by approximately 6 points per missed year of school compared with children of similar backgrounds who were in school [135]. Indeed, one of the studies demonstrating an increased IQ in the general population found that the lower-scoring segment of the population had both the greatest increase in IQ and the greatest increase in time spent in school [136]. Other possible environmental contributions to the general increase in IQ include reduced family size; higher parental education level; greater knowledge about effective parenting, including decreases in physical and highly critical disciplinary techniques; more and better children's books and toys; more reading to children; and better teachers and curricula in early elementary school programs [111,129].

Some researchers believe that intelligence has at its core a single cognitive capacity that is manifest in varying degree in all tests of intellectual ability [e.g., 74]. Others believe that there are multiple distinct components of intelligence, different types of intelligences, which can vary independently [e.g., 137, 138]. It is not known which aspect(s) of intelligence have been increasing in the general population. It is clear, however, that the increase is not due simply to practice at taking IQ tests, or increasing curricular focus in preparation for IQ testing. IQ testing has become less common in recent decades and several of the tests used to measure intelligence have been shown to be quite insensitive to the effects of practice. Moreover, the general intergenerational increase in intelligence has been demonstrated with a variety of tests that together assess a wide range of cognitive functions. Given that practicing the violin regularly alters the functional organization of the cerebral motor strip [110], the apparently environmentally induced change in the population's IQ is most likely additional evidence of the social and cultural mediation of brain function. This is so whether or not environment alters that aspect of brain function that underlies the heritable components of IQ, or closes the gap in the aspects of intelligence that distinguish an Einstein from the average citizen.

It may fairly be said that even supposedly "culture-free" tests of intelligence are products of the same culture that creates the form and content of the mass media and shapes the educational curricula, and therefore they cannot be truly culture-free measures of the effect of the culture on the brain. This, however, does not change the fact that the cultural environment alters brain function. Indeed, the measures of outcome in well-designed scientific experiments are chosen precisely because they are likely to be sensitive to the effects of the experimental manipulation being evaluated. Moreover, only from some imaginary perspective in which the brain is seen as somehow being outside of culture would it seem important, possible, or appropriate to have an objective, culture-free test of whether the brain has really changed as opposed to merely coming to function in ways that are consonant with characteristics of the prevailing cultural milieu.

Sensory Deprivation

Solitary explorers, solo long-distance sailors, survivors of shipwrecks forced to await rescue alone, and pilots or other professionals whose work requires long hours alone in monotonous sensory environments have described alterations in mood, perception, and cognition resulting from extended periods of decreased sensory stimulation. Medical researchers have noted similar effects in patients confined in bed in hospital surroundings that are almost devoid of changing or meaningful sensory input. Such effects in patients can be reversed or prevented by providing access to music, television, and interpersonal contact [139]. Stimulated by these observations, investigators have studied the effects of sensory deprivation on mood, perception, and cognition. Significant effects on each have been consistently observed.

Several methods have been used to produce sensory deprivation in the laboratory. Most commonly, subjects have been asked to lie quietly in small isolation chambers with reduced sound and controlled light. White noise may be played to mask low-level background noise and provide an unchanging auditory sensory environment. In most studies, the subjects are asked not to speak or make other noises. In some experiments the room is totally dark, while in others uniform low-level illumination is present and the subjects wear goggles that prevent perception of form. In some studies, they wear stiff, oversized gloves and cardboard cuffs around their wrists, which together limit tactile input and movement of the upper extremities. Movement may be further limited by lightly restraining the subjects or by having them lie in foam rubber cutouts that conform to their body, or in "iron lung" breathing machines that fit snugly around their body. In an alternative approach, subjects float in buoyant salt water with their ears submerged and their eyes covered with goggles.

Time in the deprivation environments has ranged from 2 hours to 7 days in different studies. After 1 hour of deprivation, most subjects begin finding the experience unpleasant. This is so even when explicit efforts have been made to create a reassuring and pleasant laboratory environment. It is so when the subjects have volunteered for the study out of an interest in experiencing sensory deprivation, have been familiarized with the laboratory setting ahead of time, and have come with plans to use the time to rest and think about a particular project or topic. J. C. Lilly, a pioneer in efforts to identify potentially positive aspects of controlled deprivation experiences, reports that by hours 2 or 3 in a flotation tank, subjects develop a "stimulus-action" hunger and show subtle methods of self-stimulation like twitching muscles, slow swimming movements which cause the water to flow by the skin, and stroking one finger with another [140].

The following is reported as a typical experience of a subject in flotation isolation for an extended period. In the second hour the subject reported a strong urge to exercise, and began whistling and singing. In the third and fourth hours the volume of his singing increased dramatically, and his songs became bawdy. The urge to action became stronger, and he reported auditory hallucinations. He was emotionally labile, crying at times, then joking, laughing, and whistling. He complained of feeling anxious and made odd statements to himself, such as "you voice, keep quiet up there, quiet." After 4.5 hours of isolation, he abruptly pulled off his mask and left the isolation tank. During the 4.5 hours, his longest period of silence had been 6 minutes. He talked for 1.5 hours during the postisolation interview, becoming increasingly angry as the interview progressed, and stating eventually, "I honestly believe, if you put a person in there, just kept him and fed him by a vein, he'd just flat die" [141, p. 542].

More objective, multisubject assessments provide a similar if less vivid account. For example, the affective responses of subjects (n = 25) kept in an iron lung in total darkness with a white noise auditory background for 7 hours were compared with subjects (n = 13) kept in the iron lung for the same period but with normal visual and auditory stimulation, and with subjects (n = 11) without either physical confinement or sensory deprivation. The group that was physically immobilized and sensory deprived described themselves significantly more often as afraid, desperate, fearful, gloomy, lonely, nervous, and panicky and significantly less often as contented, friendly, or joyful, than did either of the other groups. Both of the physically confined groups had significantly more somatic complaints than the third group, but the group that was also sensory deprived had significantly more somatic complaints than the confined but not deprived group [142]. Similar results were obtained with subjects lying on a bed in total darkness and silence compared with lying on a bed in the same isolation chamber but receiving an unusually high level of auditory and sensory stimulation, even after they had spent an initial "control" day in the same room [143]. After only 2 hours in sensory deprivation, the subjects talked for significantly longer when they described visually ambiguous stimulus cards than they did before deprivation [144].

If they are given the opportunity to receive periods of sensory stimulation during the course of sensory deprivation, subjects request such stimulation significantly more often than subjects who are not sensory deprived. The longer the period of deprivation, the more requests for sensory stimulation there are, and the more variable and complex the stimuli, the more desirable they are [145–147]. To be desired, the sensory stimuli need not, however, be of particular interest. Men isolated but not confined in a totally dark and sound-reduced room for 9.5 hours and then allowed to continue in the quiet or, by holding down a lever, to listen to recordings of stock exchange transactions, requested the auditory stimulation significantly more often than they did after spending 9.5 hours in the same room with lights on, listening to the radio, reading books or magazines, or watching television [147]. Moreover, those who rated themselves most depressed and least happy during the deprivation experience requested the most auditory stimulation. Self-ratings of fear or anger were not significantly associated with the degree of stimulus seeking, indicating that while these feelings can be prominent during sensory deprivation, perhaps especially with confinement, they are not the only, or necessarily the primary, feelings associated with the experience.

Discomfort in the absence of sensory input and the pursuit of stimulation apparently for its own sake (i.e., stimulation without extrinsic utilitarian value) has led some investigators to consider stimulus seeking an inherent drive or motivation manifest in the sense of curiosity and the experience of play [e.g., 146, 148–153]. Several studies in animals provide further support for this view. Given boards covered with easily removed red screws and unmovable green screws, monkeys soon remove all the red screws [153]. Placed in cages with shaded windows, monkeys repeatedly push buttons to open the screens, with no other apparent reward than the resulting visual stimulation itself [153]. The already impressive amount of time monkeys spend in visual exploratory activity (40%) is increased further after a period in which such behavior is restricted [154], as is the case with other appetitive behaviors.

More directly, sensory stimulation can be used instead of food as a reinforcement in learning or conditioning paradigms. Onset and/or termination of a light (i.e., a change in visual stimulation) are effective laboratory reinforcers for mice [155] and rats [156]. A monkey kept in a covered cage during training learned a color discrimination task when each correct response was reinforced with a 30-second peek through a window into the laboratory [157]. Similarly, monkeys housed in dark boxes learned to press a bar for a reward of half a second of light. Just as food is a more efficient reinforcement in hungry than in satiated animals, the rate of bar pressing increased in proportion to the length of time the monkey had been deprived of visual input [158].

In addition to discomfort, sensory deprivation in humans also produces perceptual and cognitive dysfunction. Many subjects in many studies report having visual sensations, and less frequently auditory sensations, while deprived of visual and auditory sensory input. In a large study of these phenomena, seventy-one young adults (all were high school graduates and most were college students) were asked to lie quietly in a small, sound-attenuated room for 72 hours. Black light shields that allowed the eyes to open and close but prevented visual input were taped to the face; auditory input was further attenuated by ear plugs; and the subjects wore gloves to limit tactile stimulation. The room included food, water, and toilet facilities. The subjects were instructed not to talk or make other noise. Forty-two percent of the participants reported "seeing" lights or spots; 20% "saw" shapes or forms; and 6% saw objects, people, or animals [159]. In this and other studies, many of these "visual" experiences were described as being beyond the individual's control, more vivid than usual imagery, and outside rather than inside the head, but only rarely were they experienced as real and lifelike [159,160]. Visual experiences such as these can occur within the first hour of sensory deprivation, and, according to two studies that recorded electroencephalographic (EEG) activity during the deprivation period, occur when the subjects are fully awake [161,162]. Subjects also commonly report difficulties in concentration and focused, logical thought [e.g., 152, 163]. These difficulties proved significantly more common with combined auditory-visual-tactual deprivation than with auditory-tactual or auditory deprivation, even though all three involved the same laboratory setting, social isolation, and mobility limitations [159].

Formal testing has provided confirmatory evidence of these subjective changes in perceptual and cognitive function. Immediately after extended periods of multimodality deprivation, tactile and pain sensitivity are increased, with some evidence that tactile sensitivity might increase following visual deprivation alone [e.g., 164–166]. Performance is similarly improved on simple auditory detection tasks during [167] or immediately after deprivation [168]. In contrast, performance on more complex vigilance and cognitive tasks is generally impaired. For example, in or immediately after periods of deprivation, subjects have significantly more difficulty than nondeprived comparison groups in searching for particular targets amidst distracting information [169], making words out of a set of scrambled letters [162,170], or in generating lists of words that begin with a particular letter [169]. These cognitive dysfunctions indicate that the absence of continuing activation of primary sensory areas of the brain, and therefore the absence of the cortical activation that normally fans out from these areas, leads to dysfunction of cortical areas that are important in multiple aspects of cognition. As Bruner concludes, perception and cognitive activity generally depend upon ongoing stimulation by variation in sensory input [171]. It is interesting that some of the same cognitive deficits can be produced by prolonged immobilization without sensory deprivation [172], and many of the deficits associated with sensory deprivation can be prevented if the subjects exercise during the deprivation period [173]. As Hebb suggested, anticipating both Bruner's comment and his empirical work itself, the neural organization necessary for thought and action appears to be maintained by constant sensory stimulation, some of which results from motor activity [174].

Summary and Conclusions

Current understanding of brain function emphasizes the role of integrated networks of neurons, corticocortical connections among different brain regions, and transformation of information through a series of cortical maps. The patterns of connections among cells central to all these processes develop through selective cell survival and growth and pruning of connections among cells. These physical processes are profoundly influenced by the type and amount of sensory stimulation. The survival of cells and the growth of new connections between cells depends directly upon the level of activity of the cells, which itself derives directly from sensory stimulation of the input pathway of which they are a part. The structural and functional organization of the brain can thus be substantially shaped by its environment. Empirical studies indicate that these effects extend beyond the primary sensory receptive areas and their immediate projections to reach the frontal lobes, several synapses beyond the sensory receptors themselves.

One set of studies demonstrating these effects produced alterations in brain functional organization by depriving animals of normal sensory input during the immediate postnatal period. While the effects were dramatic, the experimental alterations in sensory input were more extreme in degree, and more focused on a single sensory modality than is likely to occur naturally (e.g., suturing one or both eyes closed at birth). Other studies compared animals raised in relatively enriched environments with those reared in relatively impoverished environments, both of which were within the range of naturally occurring conditions. Again, the effects were concrete and substantial.

The human brain is particularly sensitive to environmental influence. In all animals the brain is most sensitive to the effects of sensorimotor activity between birth and the attainment of sexual maturity. This period is much longer in humans than in any other species. The effects of environmental input are greater on the cerebral cortex than on other parts of the brain. The size of the cortex relative to the rest of the brain is greater in humans than in any other species. Experimental data demonstrating the effects of sensorimotor activity on the structure and function of the human brain are available from studies of patients with early-life abnormalities of vision, studies with new brain functional imaging methods, and studies of the effect of rearing environment on intelligence.

Adult humans do not like to be without sensory stimulation, even if they have no instrumental need for information. When they are in conditions of sensory deprivation, people seek stimulation and soon become depressed and anxious. Moreover, their brains no longer work as effectively; they have illusory sensory experiences, altered perceptual thresholds, and difficulty with certain types of problem solving. This sensory dependence has just as much of a physical basis as other addictions. Moreover, this sensory dependence ties the individual to his or her sensory environment, makes it exceedingly difficult for a person to ignore continuing sensory input, and makes it extremely problematic when there are discrepancies between the new sensory information and existing structures shaped by previous environmental experience. These discrepancies are the topics of chapters 4 and 5.

The sensory stimulation considered thus far with regard to environmental shaping of brain structure and function has been primarily impersonal, for example, elimination of basic sensory information of all four modalities, conditioning rewards with light flashes or views of the laboratory, access to auditory reports of stock transactions. Although it was not the focus, control of the social dimensions of interaction has been implicit in some studies. Enriched environments in animal studies have sometimes included housing with other animals; IQ studies of only children have assessed the effect of the absence of sibling interactions; and sensory deprivation experiments have also been studies of social isolation. Interaction with other individuals of the same species is a source of environmental stimulation that has powerful and specific effects on development of the brain. The next chapter focuses on these effects.

3 Effects of the Social Environment on Brain Structure and Function

The class mammalia is distinguished by the fact that females of mammalian species nurse and care for their offspring. The name *mammalia*, based on the presence of the milk-secreting mammary glands, suggests that the provision of milk itself is the most important aspect of this process. Even initial appreciation of the broader importance of nursing-related mother–infant interactions for infant development focused on experiences of hunger frustration and satisfaction, and oral stimulation and gratification. However, the provision and consumption of milk per se are only part of a much broader, polymodal social-sensory relationship between infant and parent.

Animal Studies

The Appeal of Somatosensory Stimulation for Infants
The research of Harlow and collaborators with infant monkeys [e.g., 1] dramatically demonstrated the importance of the somatosensory relationship between mother and baby. In these studies, infants were separated from their mothers and raised in cages with access to both a wire mesh and a cloth "surrogate" mother.

Both surrogate mothers were kept at the same temperature as normal monkey mothers. Half of the monkeys received milk from the wire mesh mother and half from the cloth mother. Both groups spent much more time on the cloth than the wire mesh mother. The differential was greater by only a small amount when the cloth mother was the source of milk. The preference for the cloth mother became greater over time in both groups, the opposite of what would be expected from a food-hunger reduction conditioning model, which would predict increasing preference over time for the food-providing surrogate mother. Harlow and Mears concluded that "the disparity [in favor of selecting the cloth mother independent of which mother provides milk] is so great as to suggest that the primary function of nursing as an affectional variable is that of ensuring frequent and intimate body contact of the infant with the mother" [1, p. 108]. In other words, instead of the provision of milk being the end goal of the mother–infant interaction in and of itself, it is a means of ensuring contact between the mother and the infant because this contact is essential for development.

Oxytocin: The "Social" Neuropeptide Unique to Mammals
Real, living mothers provide a much wider range of sensory and behavioral stimulation for their offspring than did Harlow's cloth surrogate mothers [2]. These maternal behaviors are less developed or apparent prior to pregnancy and childbirth, and are initiated and maintained by a variety of changes that take place during pregnancy and childbirth, including hormonal and neurochemical changes [3]. One neuropeptide involved in this process, oxytocin, is of particular interest for three reasons.

First, oxytocin plays a role in a wide range of maternal behaviors. The initiation of maternal behavior in rats, for example, requires new mothers to overcome an aversion to the odor of neonates. At the time of birth, oxytocin is released in the olfactory area of the brain and inhibits the firing of olfactory neurons. This appears to allow a functional reorganization of olfactory response that eliminates the aversion to neonate odors [3]. In sheep, vaginocervical stimulation that mimics the effects of giving birth can induce the release of oxytocin in the brain and the initiation of maternal behavior even in nonpregnant ewes [3]. If ewes are given epidural anesthesia that prevents neural transmission from the vagina and cervix to the brain, the effects of vaginocervical stimulation on oxytocin release and ewe behavior are blocked, indicating that the effects depend on brain activity. Not surprisingly, then, oxytocin injected into the brain can stimulate maternal behavior in ewes who give birth with epidural anesthesia and in nonpregnant ewes [3].

The second thing of interest about oxytocin is its role in social behavior more broadly. Approximately 5% of mammals form stable parental dyads [4, 5]. In monogamous prairie voles, parental pair bonding can be facilitated by injection of oxytocin or another similar neuropeptide called vasopressin [3]. Conversely, behavioral aspects of monogamy can be inhibited by injection of compounds that block oxytocin or vasopressin receptors [3]. Both oxytocin and vasopressin are released when prairie voles mate, thereby promoting the development of stable social bonds between future parents [3]. In mice, oxytocin appears to be necessary for adult social learning and recognition of familiar conspecifics. When the gene that directs synthesis of oxytocin is knocked out, and the mice are unable to make oxytocin, they are completely unable to recognize other mice, even after repeated encounters with them [6]. Other types of memory seem to be totally normal. Moreover, if oxytocin is injected into a region of the brain called the amygdala, social learning is restored [6]. Brain functional imaging studies in human beings show that the amygdala becomes active when subjects look at emotionally expressive pictures of peoples' faces [7, 8].

The third and most interesting thing about this remarkably versatile neuropeptide is that oxytocin is found only in mammals [3]. It is thus a concrete marker of the defining importance of social interaction in the development and life of mammals.

Effects of Decreased Maternal Stimulation on Infant Development

The critical importance of polymodal interactions between parent and infant for normal development of the infant has been demonstrated in many experimental studies in animals. For example, when 10-day-old rat pups are separated from their mothers for only 1 hour, growth hormone levels in the blood drop to half of their normal value [9]. Similarly, ornithine decarboxylase (ODC) activity, an enzyme important in protein synthesis and an index of tissue growth and differentiation, drops to half-normal levels in the brain and other body organs. After the pups are returned to their mothers, values first rise substantially above normal and then, within an hour, return to normal. If while separated from their mothers the pups are stroked with a camel's hair brush to simulate licking by their mother, growth hormone levels and ODC activity both remain normal. Other types of sensory stimulation, including rocking, passive movement of limbs, and tail pinching fail to provide similar results, demonstrating the specific importance of licking for the pups. Indeed, when pups remain with their littermates, and with a mother who has been anesthetized but remains available for nursing, warmth, and olfactory stimulation, levels drop as if the pups were separated from their mothers. Thus, even access to nursing cannot maintain the physiological processes essential for normal growth in the absence of other active, specific, sensory stimulation by the mother. The relative unimportance of nursing itself for maintaining growth hormone levels and ODC activity is further demonstrated by measurement of ODC activity in rat pups kept with a mother whose nipples have been ligated. Despite the fact that the pups were accustomed to feeding approximately every 10 minutes, the experimentally created milk deprivation did not lead to the same drop in ODC activity seen in either pups separated from their mothers or pups kept with anesthetized mothers.

The dependence of infant growth hormone levels on specific contact with the mother is compounded by a similar dependence of tissue sensitivity to growth hormone. Growth hormone regulates a variety of physiological processes throughout the body, and injection of exogenous growth hormone in normally reared rat pups leads to a fivefold increase in ODC activity. However, if the injection is given after 2 hours of separation from the mother, it has no effect on ODC activity. Thus, in addition to decreasing growth hormone levels and ODC activity, separation seems to decrease tissue sensitivity to growth hormone.

Similar processes are probably responsible for the pathologically diminished growth seen in institutionalized human infants who are provided with adequate food but not with normal social interaction [e.g., 10–12]. Indeed, extremely premature low-birthweight human neonates (mean gestational age 31 weeks, mean birth weight 1,274 g) who were given extra tactile and kinesthetic stimulation while in the hospital (45 minutes extra per day in this study) grew more quickly (47% greater weight gain), spent more time awake and alert, reached developmental milestones more quickly, and were discharged home from the hospital sooner than a comparison group who received standard hospital care [13, 14]. Moreover, the experimental group grew more than the control group despite the fact that both groups fed equally often and consumed similar amounts of food.

The effects of maternal separation on growth are also evident in the brain itself. Rat pups separated from their mothers for 24 hours showed a twofold increase in the death rate of neurons and glial cells in the cerebral and cerebellar cortices, and in the white matter tracts that link different brain regions [15]. Monkeys repeatedly separated from their mothers for 5-hour periods between the ages of 13 and 21 weeks had abnormally large right frontal brain regions as adults. Although the time at which this subsequent abnormality developed is unclear, it is thought to probably result from an abnormal decrease in the cell pruning that usually takes place around the time of sexual maturation [16].

Lasting Effects of Maternal Stimulation during Infancy on Adult Brains and Behavior

Neurochemical studies of adult animals have found persistent abnormalities in multiple neurotransmitter systems in animals separated from their mothers during infancy. Expression of the dopamine transporter gene and dopamine-mediated stress responses [17], expression of serotonin receptor mRNA [18], expression of benzodiazepine receptors [19], sensitivity of glucocortocoid receptors related to stress response [20], and sensitivity to morphine [21] are all altered in adult rats who have been separated from their mothers for varying periods of time as infants. Experimentally induced autoimmune encephalitis is more severe in adult rats that had been separated from their mothers as infants, suggesting an altered immune system function [22].

The actual neurochemical mechanisms through which early life experiences can affect neurochemistry and behavior in adult life have been identified in a series of studies of rat pups (23, reviewed in Ref. 24). It has been known since studies in the 1960s that separating pups from their mothers for 3–15 minutes every day leads to decreased stress reactivity, decreased fearfulness, decreased steroid release from the adrenal glands during stress, and decreased aging of the hippocampus when the pups are adults. The story became more interesting when it was noted that after the rat pups were returned to their cages following the short periods of separation, their mothers increased the amount of time they spent licking and grooming the pups. Perhaps it was this increased maternal attention that led to the decreased stress reactivity and other behavioral changes in these rats when they became adults. Even without experimental manipulations such as the short separations of pups from their mothers, there is considerable variation from mother to mother in the amount of time they spend licking their pups. This made it possible to study the neural mechanisms through which early-life sensory experience might produce lasting effects on brain function by comparing offspring of mothers with naturally occurring differences in maternal behavior.

Following this strategy, Michael Meaney and colleagues found that adult rats who had been licked more as pups had stress responses similar to rats who had been separated from their mothers each day (and then licked more upon return). This showed that the effects observed in the "separation" experiments were operative within the range of variation in naturally occurring rat behavior and were not limited to more extreme situations created by researchers [23, 24]. It remained possible, however, that the differences in stress response between adult rats who had been licked more or less as pups were characteristics they inherited from their mothers rather than characteristics they developed as a result of postnatal experience. To distinguish between these two possibilities, Meaney and colleagues took pups born to high-licking mothers and immediately after birth placed them with low-licking mothers. They also did the reverse, taking pups born to low-licking mothers and placed them with high-licking mothers. When these rats became adults, their stress responses were consistent with the type of mother that reared them and not with the type of their biological mother.

Further studies identified actual changes in the genes associated with stress response as a result of the degree of maternal licking. Shortly after birth, the surface of DNA is largely covered by small chemical complexes called methyl groups. These methyl groups limit access to the DNA and thereby limit activation or expression of genes. Experiences during the first weeks of life can lead to selective removal of these methyl groups, making some genes more active. The effects of experience on methylation are much greater during the first 3 weeks of a rat's life than thereafter, and changes induced by experience during this "critical" period remain relatively unaltered throughout the rat's adult life. By again using cross-fostering experiments, Meaney and colleagues showed that maternal licking initiates a series of neurochemical processes that selectively demethylate the genes that produce the glucocorticoid receptors in the hippocampus and frontal lobes that turn off the stress response. Through these processes, the early-life sensory experience of being licked and groomed leads to lasting changes in the structure and function of genes that regulate response to stress.

Some of the persistent neurochemical and behavioral effects of maternal care of female infants affect the way the infant functions as a mother herself when she becomes an adult. Females that had been separated from their mothers when they were infants showed lower than normal gene expression in areas of the brain associated with maternal behaviors when they themselves became mothers [25]. They also licked and crouched over their pups less often than other mothers [26], and their generally decreased ability to maintain attention and their increased response to stress have been hypothesized to further compromise their maternal competence [25]. Such intergenerational effects are potentially self-propagating and even self-amplifying. Moreover, since litter size [25, 27] and food availability [16] can influence the amount of licking and other behavioral interactions between mother and infant, a variety of environmental factors can influence maternal behaviors and their impact, across generations, on a range of individual and group behaviors.

Experimental studies in monkeys have demonstrated the effects of maternal deprivation on more complex aspects of adult social behavior, and some of these effects seem to have direct parallels in human beings. Comparisons between infants separated from their mothers shortly after birth and raised in small peer groups and infants reared by their mothers are most informative since both groups have ample social and physical contact with members of their own species. The major difference between groups is the presence or absence of a biobehaviorally mature mother. Peer-reared and mother-reared monkeys show many behavioral differences even after reaching maturity (reviewed in Ref. 28). As infants and juveniles, peer-reared monkeys spend much more time clinging to one another and sucking on their own fingers and toes. Interactions among peer-reared monkeys appear chaotic and are characterized by rapid fluctuations between periods of isolation and intense engagement. Aggressive behavior is less common in peer-reared animals but more often leads to injury. Peer-reared juvenile monkeys show more signs of stress when they are separated from their cage mates and are rated by trained observers as generally more tense, timid, and emotionally labile. Peer-reared and mother-reared monkeys differ in both baseline and stress-related levels of brain neurotransmitters and circulating hormones, with the peer-reared monkeys showing blunted responses in some and exaggerated responses in others.

Together these observations indicate deficient biobehavioral self-regulation in the peer-reared monkeys. Additional examples of deficient self-regulation are found in monkeys reared in partial or total isolation. Such individuals show altered temperature regulation, abnormal eating patterns marked by polydipsia and hyperphagia, and impaired regulation of body weight [29]. Consistent with this, monkey infants reared by inanimate, unreliable, or abusive mothers generally show more severe responses to separation than do monkeys reared by real and competent mothers [28]. The inanimate or inadequate mothers have less survival value than the highly competent mothers, and if the degree of separation distress were based upon the value of the absent object rather than the efficacy of self-regulatory mechanisms that restore equilibrium, one would expect infants reared by competent mothers to experience greater distress. Similarly, human infants whose mothers were rated by researchers as less sensitive and responsive to signals from their infant showed more negative responses to being put down, had longer periods of crying in general, and cried more when their mothers left the room [30].

Beyond deficits in self-regulation, peer-reared monkeys show evidence of decreased neurobiological structure or organization. Mother-reared monkeys show significant correlations among the activity of different brain neurotransmitter systems, among different measures of behavior, and between neurotransmitter and behavioral measures. Most of these correlations are significantly reduced or absent in peer-reared monkeys [28]. A similar loss of neurobiological structure has been reported in human infants raised under conditions of near social isolation in institutional nurseries. In addition to a lack of facial or vocal expression, social withdrawal, and bizarre stereotypic movements similar to those described in socially deprived monkeys, these infants failed to show the normal daily cycles in cortisol level [31].

In an effort to make more ecologically natural experimental interventions in maternal–infant interactions, several experiments have evaluated the effects of requiring mothers to spend more or less time foraging for food [e.g., 16, 32]. These variations in maternal availability within the range of what could be expected to occur naturally had effects on the development and behavior of offspring that were large enough for researchers to be able detect them with the relatively crude measures of brain structure and function now available. In one study, a third experimental group was created in which for half of the rearing period mothers were in a low-foraging environment and for half they were in a high-foraging environment [32]. This mixed condition had a greater effect on infant development than did the consistently high-foraging environment. As discussed extensively in chapters 4 and 5, it is problematic for environmentally dependent organisms when the environment (in this case, maternal presence and behaviors) changes so that it no longer matches the expectations and internal neuropsychological structures generated by the previous environmental conditions.

Human Studies

Parental and social behaviors vary widely among mammals, as do the neurochemical systems related to these behaviors. The role of oxytocin in social learning among rats appears to be quite different from that in mice [6, 33], and while human beings do have oxytocin receptors in their brains, the receptors are in different brain regions than in monogamous voles and monkeys [3]. Behaviorally, visual and auditory interactions play a much more central role in human infant–parent bonding and social connections more broadly, and olfactory input is relatively less important than it is in many other mammals.

Within the first 2 or 3 days of birth, human infants are able to distinguish their mother's voice from that of other women and turn more readily to it [34, 35]. They are already familiar with their mother's voice because it is one of the most intense acoustical signals measurable in the amniotic environment [35], and they can even distinguish the language spoken by their mother during pregnancy from other languages [36]. Such prenatal familiarity is then complemented by extensive early postnatal exposure. Thus, during an extended period of pre- and postnatal sensory-dependent development of the auditory cortex, the mother's voice is the single most consistent stimulus.

By 20 days of age infants will suck significantly more on a nipple in order to hear a recording of their mother's voice than they will to hear the voice of a female stranger [37].

Similar early sensory sensitivity to social stimuli is also evident in the visual system. Within hours of birth, human infants show selective visual sensitivity to the human face, looking at and following with their eyes schematic drawings of a face significantly longer than other drawings similar to faces in the number and nature of individual visual elements [38]. This apparently inborn interest in human faces leads the infant to attend selectively to this component of its visual environment. By 14 days of age, human infants prefer looking at their mother's face rather than the face of a woman who is a stranger to them [39]. Such preferences soon extend to others whom they also see often. The infant's preference for faces in full front presentation as opposed to faces seen in profile [e.g., 40] contributes to the infant and the adult simultaneously focusing attention on one another.

These innate social-sensory predilections, the dependence of growth itself upon social-sensory stimulation, and the intensity of pre- and postnatal social-sensory experience, rapidly link the infant and primary caregiving adults in mutually regulating dyadic systems (reviewed in Refs. 41, 42). The resulting synchronization and mutual contingency of infant–adult behaviors are evident in sleep and electroencephalogram patterns [43], direction of gaze [44, 45], facial expressions [46–48], vocalization [49–51], and cardiac and behavioral rhythms [52].

The activity of the adults who are thus linked with infants affects the infant's general physiological state and orientation to sensory input. Rocking, walking, and other kinesthetic stimulation help quiet agitated babies and ease the transition to sleep [e.g., 53–55]. Slow, deep vocalization helps calm an agitated infant, an increased pitch and timbre of the adult voice can rouse a baby from light sleep to alert awareness, and increased tempo and a staccato rhythm can overarouse an infant to the point of crying [56]. Infants even move in synchrony with the speech rhythms of the adult interacting with them [57].

Such connections with an infant produce a variety of psychobiological changes in parents [e.g., 58–60], and the biobehavioral parallelisms and reciprocities evident between infant and parent result in part from these changes in the parent. However, it is the developing and relatively undifferentiated infant that is the more affected as it grows linked with the mature, structured, and self-regulating parental component of the dyadic system.

Infancy and childhood last much longer in humans than in other mammals, allowing greater influence of these social interactions on brain development. The processes through which such interactions have their effects have been well studied, albeit most often descriptively rather than experimentally. Several distinct but overlapping and interacting processes have been described: instrumental parenting, turn taking, imitation, identification and internalization, and play. These processes are the basis for the long-lasting effects of the social environment on development of the human brain, the sensitivity of humans to change in their environment in general and their social environment in particular, and the great efforts humans will make to maintain constancy in their environments.

Instrumental Parenting: Regulation and Training of Infant Physiology

The most varied of the processes, instrumental parenting, includes adjustment and training of infant physiology, creation of the physical environment in which the infant and child develop, intervention in object-directed activities, and provision of motor and cognitive functions not yet developed in the infant. Adults regulate the infant's state of arousal by speaking in calming or exciting ways; by positioning the baby in supine or upright positions; by holding the baby to the breast so it can suck, feed, feel the warmth and tactile stimulations of contact with the mother's skin and learn the familiar sounds emanating from the mother's heart and lungs; and by touching and moving the baby in repetitive and familiar ways. The same procedures help infants return to equilibrium after painful or otherwise distressing events. Adults come to recognize early signs of infant distress and respond with comforting interventions. Such interventions, of course, work by inducing neural activity in the infant that leads to reestablishment of equilibrium. As particular parental responses are repeated over time, especially when the infant is in similar neurobiological states of early distress, the quieting neural responses in the infant become quicker and stronger. Since the quieting responses are consistently linked with the onset of distress, the onset of distress can itself become an elicitor of the neural processes that contain it,' and self-regulation develops.

The presence of adult-generated sensory input, through a soothing voice, calming touch, or rocking, for example, further enhances the quieting neural processes intrinsic and developing in the infant, and if the stress is great enough, such input may be essential for restoration of equilibrium. The more familiar the adult input and the more similar in detail it is to that provided on past occasions, the more effective it is. Thus, an infant is often quieted more quickly when it is held and soothed by its mother than when it is held and soothed by another woman, even if the second woman has had as much or more experience caring for infants as the infant's own mother. Adult sensitivity to early signs of infant distress or other changes in state comes from repeated observation of an infant's responses, and much is apparently learned without conscious effort or even awareness. Awareness can come from the parents empathically noting changes in themselves that result from changes in state in the infant half of their dyadic systems. Discussion with other adults caring for the infant, with adults experienced in the care of other infants, and with individuals designated as experts by the larger social group further increase explicit awareness of the signs of an infant's distress. Parental modulation and training of infant physiology extend beyond the dimensions of sleep-arousal and equilibrium-distress and include eating behaviors, eliminative functions, and exploratory behaviors [e.g., 61]. Linked with their parents in dyadic and family systems, infants develop physiological patterning that is influenced by and often similar to the patterns of their parents.

Instrumental Parenting: Creating New Objects in the Rearing Environment

A second aspect of instrumental parenting is adult intervention in the infant and child's object-directed activities. Kaye observed that "social interference in the object-directed activities of babies is such a commonplace occurrence that few authors have remarked on its absolute uniqueness to our own species" [62, p. 193]. First, only humans create a large part of the inanimate environment that is perceived and acted with and upon by their offspring. Use of these human-made objects by infants and children alters their psychobiological development. The development of visually directed behavior in infants is accelerated by hanging objects of visual interest in their cribs [63]. As discussed earlier, new brain imaging techniques have shown that socially endorsed, repetitive use of human-made objects such as musical instruments is associated with actual changes in brain structure [64]. As discussed in chapter 2, school, perhaps largely through use of that most influential of human inventions, the printed word, appears to increase brain functional capability as measured by intelligence tests [65], and the originally robust but now decreasing difference in IQ between rural and urban populations has been attributed to the degree of exposure of rural residents to the objects and complexities of the human-made environment [66]. Increasingly, children develop in a human-made rather than a naturally occurring environment.

Instrumental Parenting: Directing and Shaping Attention

Second, adults intervene in a child's object-directed activities by directing the child's attention from one object to the next and thereby influence not only the particulars of such activities but also the development of the child's own attention-directing processes. If an infant is given a choice of playing with an object being handled by an adult or with an identical copy of the object that is closer, the infant will reach past the copy to play with the one the adult has [67]. By 6 months of age more than half of infants will follow their mother's gaze, and by the time they are a year old nearly all will do so [68–70]. By 9 months, most infants point to objects [71], and at this same stage in development, parents increase their use of pointing when interacting with their infants [72]. Initially infants respond to movement of an adult's head in one direction or another, but by 18 months of age they are able to follow eye movements as well [70, 73]. The ubiquity of these mechanisms of shared attention has led some to suggest that they are inborn, but experiments have shown that infants who are only 6 months old and do not initially follow an adult's gaze in the laboratory can learn to do so through simple conditioning [70]. Thus, even if there is an inborn component of this important social developmental mechanism, the mechanism itself can be further affected by social interaction. These mechanisms permit relatively rapid and selective direction of attention to objects at a distance as well as those close at hand, and allow parental influence over infant perceptual (and related cognitive) activity as well as motor activity. Through such means, adults influence what in the continuous stream of sensory input infants are most aware of, become most familiar with, and think most about.

The corresponding effects on brain activity are pronounced, making internally concrete the invisible connection between a pointed finger and an attended object. Recordings of brain electrical activity from the scalp or from the surface of the brain itself have identified specific wave forms associated with perception, reflecting the sequential activity of different groups of cells involved in different stages of sensory processing. When attention is directed toward a stimulus, the amplitude of the deflection increases, as does the extent of the scalp or brain over which they can be detected, while if attention is directed away, there are corresponding decreases (reviewed in Ref. 74, pp. 265–268). For example, potentials evoked by presentations of sounds to a cat are markedly decreased when the sounds are presented at the same time that the cat can smell or see a mouse [75].

Functional magnetic resonance imaging (MRI) studies in humans provide similar evidence. For example, if subjects attend primarily to the sound track of a movie rather than the visual portion, activity in the auditory processing areas in the temporal lobes is marked and activity in the visual areas of the occipital lobes is limited, even though both auditory and visual information is reaching the brain. When they switch attention from the audio to the visual portion of the movie, activity decreases in the temporal lobes and increases in the occipital lobes. Such effects are even evident within the visual field itself. If subjects attend selectively to one half of the visual field because that is where "target" stimuli occasionally appear, brain activity is much greater in the corresponding area of the visual cortex, even though most of the visual stimulation is distributed evenly throughout the entire visual field [76]. Given the critical importance of neuronal activity in establishing and maintaining interneuronal connections, the influence of adults on the attentional mechanisms that can so markedly and selectively amplify some neuronal activities and diminish others must shape neuronal circuitry in the developing brain of the infant and child. This circuitry is shaped by what the adults are interested in and direct the infant's attention toward, features which themselves are the product of the adult's own childhood experiences and the continuing effects of their adult social community.

Writing before the just-cited studies that provide support for their views, Vygotsky and Luria argued that parental direction of an infant's attention is the origin of the child's ability to direct its own attention [74, 77]. To quote Vygotsky:

In the early stages of development the complex psychological function *was shared between two persons*: the adult *triggered* the psychological process by naming the object or by pointing to it; the child *responded* to this signal and picked out the named object either by fixing it with his eye or by holding it with his hand. In the subsequent stages of development this socially organized process becomes reorganized. The child himself learns to speak. He can now name the object *himself,* and by naming the object himself he distinguishes it from the rest of the environment, and thus directs his attention to it. The function which hitherto was shared between two people now becomes a method of *internal organization of the psychological process.* From an external, socially organized attention develops the *child's voluntary attention,* which in this stage is an internal, self-regulating process [quoted in 74, p. 262].

Thus, from a very early age, social interaction with adults shapes the mechanisms that underlie social interaction, externally influences the direction and manner in which the infant's attention is deployed, and shapes the development of the self-regulatory mechanisms that direct attention throughout the individual's adult life. As a result, the self-regulation of attention varies in its particulars from person to person, culture to culture, and historical moment to historical moment.

Instrumental Parenting: Language and Other Symbolic Media

Language allows still more powerful, extended, and broad social influences to operate on the direction of attention and the development of voluntary attentional capacities. First, it allows finer distinctions in shared direction of attention than indicated by a gaze or pointing. Second, it allows the direction of ongoing internal cognitive and retentive operations after the original event or object of perceptual attention is gone. Third, it permits individuals who were not and are not even in the same place as the child whose attention is being directed to exert an influence. Given its powerful role in social mechanisms that help develop an individual's ability to voluntarily direct their own attention, it is not surprising that language seems to play an important role in these human abilities once it is established.

Vygotsky emphasizes the importance of silently naming an object in order to focus one's attention on it. Experimental studies confirm that naming an object can increase the sensitivity of sensory processing, as could be expected from the evoked-potential and functional brain imaging data cited earlier that show increased neuronal activity when attention is focused on an object that is already in the sensory field. Beyond its role in self-direction of attention, inner language facilitates many other cognitive operations; for example, providing a vocabulary and structure for thought itself, making all such operations dependent upon this most human and most interpersonal of capacities.

It is known from the effects of localized brain lesions on behavior, and has recently been confirmed in several brain imaging studies [e.g., 78], that late-maturing frontal lobe regions play critical roles in language function in general and word generation in particular. These same regions are critical in the voluntary direction of attention. For example, animals that have had their frontal lobes surgically removed cannot inhibit automatic responses to irrelevant stimuli and thus are unable to sustain goal-directed activity or perform short-term memory tasks. If, however, an animal is placed in complete darkness or given tranquilizers that lower its general sensitivity to sensory input, its performance on memory tasks improves [71, pp. 90–91]. Thus we see that language-mediated, socially influenced regulatory processes are subserved by those regions of the brain that most distinguish the human brain from those of other mammals and that have the longest period of postnatal growth and development. These processes regulate the activity of the brain itself, and since the brain's activity shapes its development, adult participation in these processes throughout infancy and childhood shapes the child's developing brain. Social influence on development of the self-regulatory function itself leaves a lasting impact on brain growth and function. Moreover, as will be discussed more fully in the next chapter, even when fully developed, the self-regulatory, self-direction of attention is far from autonomous, and social input remains a built-in feature of these mechanisms throughout the life-span.

In recent years the rapid development and dissemination of largely nonverbal electronic games, which are often interactive, has created a new and fundamentally different mode of social shaping of brain development. Designed to capture and hold children's attention, these games can provide hours per day of perceptual, motor, and cognitive exercises. Although their effect is not yet well studied, early reports document the influence of these games on the development of cognitive skill [79]. Efficacious therapeutic exercises have been developed that are aimed explicitly at inducing functional reorganization of the brain and enhancing specific underfunctioning cognitive and perceptual systems through sustained, repetitive performance of gradually and progressively more difficult variations of the same activity [80–82]. There can be little doubt that spending many hours a week of childhood on similar recreational games creates additional and perhaps even more powerful opportunities for social shaping of the particulars of brain development. Some observers have noted that deployment of attention is markedly different while playing most electronic games or watching many new music videos than during most preelectronic activities. Most notably, attentional shifts are rapid from item to item within the general activity, while attention to the activity itself is intense and sustained. Moreover, in most electronic games, language is absent and the speed of processing makes much use of inner language and labeling counterproductive. In many music videos, when language is present the words are unintelligible. The language system is thus used in a different and subordinate manner, compared with its use in reading or in internal or external speech. These new, electronic activities have the potential to substantially alter the functional configuration of adult human brains.

Instrumental Parenting: Provision of Motor and Cognitive Functions

In addition to creating much of the child's environment of objects, and directing its attention from object to object or activity to activity, the instrumental parent participates with the child in shared activities, providing components of motor and cognitive functions of which the child is not yet capable. These functions include setting goals, selecting strategy, collecting necessary tools or objects, focusing and sustaining attention, and executing fine motor aspects of the task. For example, a parent may place before an active infant a sorting box with holes of different shapes on its top and objects of corresponding size and shape around it. After the infant picks up an object, the parent may guide its hand to the appropriate opening and help push it through. Later, if the infant carries the object to an appropriate opening, the parent may guide the infant by hand through the fine motor adjustments necessary to get the object through the hole. Through facial and vocal expressions, the parent may direct the infant away from an opening that does not match the object in its hand to one that does. When all the objects are in the box, the parent empties the box so the activity can begin again. Operating thus within a structure provided by the mind of the adult, i.e., a larger schema that integrates a series of component sensorimotor routines into a larger behavior, the infant can practice and develop sensorimotor routines that in themselves are of limited functional value and without direct reward but that become of great value when they are integrated with routines the infant will not master until later. The knowledge of this value is in the mind of the adult, who brings the infant's incomplete efforts to fruitful reward or who directly rewards the infant's partial accomplishments themself.

As the infant becomes proficient at an activity, the parent provides new variants of increasing difficulty; new boxes, for example, with more numerous and more finely discriminated openings and objects. As a child develops increasing competence in an activity that ranges broadly in difficulty, it is the adult's memory that keeps track of the child's performance, thus allowing the adult to renew the activity at progressively more demanding but not too demanding levels, and to keep the infant or child functioning in what Vygotsky called its individual and constantly changing "zone of proximal development." And it is the adult's memory that first allows an infant or child to stop a task when further effort at the time is not possible and return to the task at a later time.

These contributions of instrumental parenting to the child's development are even more vividly evident in neurobiological terms. Most of the component functions provided by the parent as he or she participates with the child in progressively more complex behaviors are functions to which the frontal lobes of the brain make primary and essential contributions. Following lesions to the frontal lobes, for example, attention, memory, organization, planning, and strategy selection are markedly impaired. Given the prolonged postnatal physical maturation of these structures in human beings, lasting until or beyond puberty, it is not surprising that adults must provide these functions if they are to be present in the behavior of infants and children. Essentially, then, the frontal lobes of parents are functionally linked with the lower brain centers and the sensory, motor, and association cortices of their infants and children. While the child's frontal lobes are developing, the parents' brains provide frontal lobe functions for the child. As discussed in chapter 2, neural activity in the primary and secondary sensory areas of the brain is propagated to the frontal and parietal lobes, and is necessary for and influential in the functional development of these brain regions. Thus, the frontal lobe activity of parents shapes the development of their children's frontal lobes through the parents' use of their own frontal lobes to structure the activity of their children's motor and sensory cortices.

Instrumental Parenting: Research Studies

Research studies on the effects of instrumental parenting have taken several different approaches. One generation of studies sought to demonstrate differences among cultures in the rate and extent to which individuals achieved the stages of cognitive capability described by Piaget (see Ref. 83 for discussion). The sequence of cognitive development described by Piaget was thought to reflect an inborn biological developmental thrust that might be altered by education and cultural context. These efforts were confounded, however, by the fact that the tests used turned out to be culturally specific rather than universal neurocognitive yardsticks. Individuals would show a particular cognitive capability on one test but not on another, while the second test would elicit the expected cognitive function in members of a second culture. Culture thus had profound effects on the abilities of people to recognize, understand, approach, and perform tasks. The last part of the next chapter presents firsthand accounts by immigrants of the difficulty they have thinking and functioning in a new cultural environment. These difficulties arise in part from the fact that the cognitive operations necessary for solving certain problems are activated in culturally specific ways. It may not be possible in a foreign culture to recognize the nature of the problem with which one is faced, or to think about the problem in the appropriate way.

A second series of experiments has demonstrated the impact of English language schooling on thinking and behavior in cultures where people did not previously speak English or go to school [see 84]. When schools were first introduced, some children in a village attended while others did not. In other instances, schools were established in one village, but not in a nearby village that shared the same culture. In both situations schooling had large effects on problem solving and the interpersonal communication of information, both in laboratory tests and tests designed to evaluate commonly occurring situations in the local culture. Indeed, the effects of schooling were greater than the effects of age; younger children who went to school outperformed older children who did not. These studies all have the virtue of comparing individuals from the same culture with measures of function derived from that culture. A final group of studies have identified specific cognitive strengths associated with particularly prominent features of individual cultures [e.g., 84]. Learned men from a nonliterate New Guinea society reportedly are able to carry 10,000–20,000 names around in their heads [84]!

Turn Taking

Kaye has made the intriguing suggestion that human infants are the only mammalian infants who nurse in bursts, sucking for 10 seconds or so and then pausing with the nipple still in the mouth, rather than sucking for a more extended period, dropping the nipple, and attending elsewhere [41]. The maternal response to this pause, he goes on to suggest, is automatic and universal. Even first-time mothers who report never before having even held a baby jiggle a baby while nursing it, especially when the baby pauses from sucking. According to his observations, infants are more likely to resume sucking just after jiggling stops than they are either during jiggling or if they had not been jiggled. When babies are 2 days old, each jiggling lasts an average of 3.1 seconds. When they are 2 weeks old, their mothers have shortened the average jiggle to 1.8 seconds. The mother's actions serve to organize this most basic and repetitive interaction between infant and mother, providing the prototype, Kaye argues, for turn-taking interactions in general.

Similar maternal interventions soon thereafter structure into interactive bursts the infant's smiles, vocalizations, and wide-mouthed expressions that initially are themselves randomly distributed in time [41]. While turn taking may not be unique to human beings, nowhere else in the animal world is it as developed, and in no other animals is it sustained over time in anything like a dialogue. Kaye remarks on the complaints of scientists who first taught apes to use sign language. The apes were "terrible about interrupting," found it difficult to take turns, would usually follow each sign with some kind of action, and could not create even a briefly sustained dialogue [e.g., 85].

With the development of language, turn taking is further refined and more firmly established. However, adults continue to play critical roles in creating, sustaining, and shaping dialogues with their children, again providing skills that their children have yet to develop. Kaye videotaped mothers sitting at the kitchen table with their 2-year-old child looking together at a picture book, playing with a toy tea set, and playing with a family of dolls and their furniture. More than 20,000 "turns" were categorized as simple responses to a partner's previous turn, as turns which in adult discourse it would be rude not to react to in some way, and/or as a combination of the two ("turnabouts"). The children produced many responses and some turns that elicited responses, but very few turnabouts. Mothers produce many more turnabouts and turns that elicit responses. This contrasts with child–child interactions, which have few turnabouts and as a result much shorter conversational chains. It also differs from normal adult–adult conversations, which also have few turnabouts but many responses that create an extended dialogue. Moreover, the mothers maintain a topic that is of interest to the child, but it never goes the other way around. In addition to choosing and/or maintaining a topic of interest to the child, the mother also provides the sustaining logic and memory of the extended conversation, including a hierarchical structure of major and subordinate themes. As Kaye summarizes, "the effect of this enduring discourse frame is to involve the infant in dialogue that is always beyond his own capabilities for intentional discourse" [41, pp.102–103].

As described in relation to the adults' role in directing and sustaining attention, these parental functions—initiation, sustaining focused attention, keeping the topic in mind, and creating and clarifying a logical structure—are subserved largely by the frontal lobes of the brain. Again, then, the parental frontal lobes are driving the infants' language and sensorimotor activities, filling in form and shaping the development of the child's own frontal lobes. As indicated by Vygotsky, what goes on as an interpersonal dialogue for years during a child's development becomes an intrapersonal aspect of the child's own thought processes. Again, the role of language is clear, first as the medium of the interpersonal process and then as the medium of thought within the individual's mind.

Imitation

Because of its ubiquity and automaticity, imitation is an important process through which human adults shape the developing brains of their offspring. Although not unique to humans, as observation of apes in metropolitan zoos attests, it is much more readily and more often elicited in human beings [86]. Moreover, the prolonged and close bond between human parent and child, and the extended period of postnatal development of the human brain, allow imitation to have a more extensive impact on human growth and development than in other primates. And although imitation does not depend on turn taking and dialogic interactions, these allow imitation of more complex and sequential behaviors. Imitation is as basic a process as the stimulus seeking or curiosity described in chapter 2. It may be understood as "structure seeking" or a search for organizational templates. It is a direct and concrete mechanism to shape the infant brain according to the particular mix of adult features, structures, and behaviors surrounding the infant on a regular basis. It has been suggested that autism, a pervasive developmental disorder that broadly compromises intellectual and social function, derives from a fundamental inability to imitate [86].

Studies in monkeys have identified cells in the frontal lobe that increase their firing activity both when the monkey makes a particular movement and when the monkey observes the experimenter making the same movement [87,88]. These "mirror neurons" are concrete evidence of the link between imitation and behavior. Brain functional imaging studies in human beings similarly show that specific areas within the frontal and parietal lobes that become active when subjects perform patterned finger movements are also active when they observe another person making the same movements. The activation is greatest when the subjects make and observe the movement at the same time, directly showing the effect of imitation on the neural system that performs the observed behavior [89]. Moreover, the areas of the brain involved, the frontal and parietal lobes, are precisely those that show the longest period of postnatal maturation and the greatest increase in size in human beings compared with other mammals. This association between the frontal lobes and imitation is further supported by the clinical observation that damage to the left frontal lobe can reduce a person's ability to imitate [90,91]. The area within the frontal lobe that is particularly active during brain imaging studies of imitation, the pars opercularis or Broca's area, is also the area that usually plays a critical role in the language function, providing a physiological basis for the close connection of these two important human functions [92].

Imitation has been demonstrated in neonates as young as 2 weeks of age. For example, one group of researchers videotaped babies for 150 seconds following one of three different adult behaviors, each lasting 15 seconds: (1) the adult held their face still; (2) the adult stuck their tongue out repeatedly; or (3) the adult opened and closed their mouth repeatedly. The infants stuck their tongues out more often after watching the adults stick their tongues out, and opened their mouths more often after watching the adults open their mouths [93]. Systematic observations of mothers and 3-month-old infants found that both were more likely to vocalize if the other was vocalizing [94]. The dramatic utility of imitation is evident in a laboratory study of 6-month-old infants. The infants were seated on their mothers' laps, facing a table on which there was a toy directly in front of them but behind a clear Plexiglas wall. None of the infants reached around the Plexiglas and got the toy themselves even when given ample time to do so. However, if the mothers simply reached around and got the toy while the infant was watching, without touching, enticing, encouraging, or reinforcing their infant's efforts in any way, all infants began within minutes to reach around the wall and grab toys [41, pp.169–170]. They were able to transfer the skill to their other hand, and to retrieve toys different from the one their mother got.

Imitation has such great developmental impact because it is consistently operative throughout the moment-to-moment unfolding of everyday life. The infant's caregivers demonstrate and model ways of doing many things many times in the course of caring for the infant and while working within view of the infant. If an infant reaches for a toy with a closed hand and as a result pushes it away, the adult returns the toy by reaching with an open hand. If a child fails to visually identify the most effective place on which to grasp an object, or fails to integrate that visual information with its motor activity and as a result drops the object, the caregiver demonstrates the correct way to obtain the object. When an infant imitates the correct way to do something, the event is often immediately noted by the adult with vocalizations of pleasure and approval. When mothers deliberately demonstrate things to their infants, most mothers move their heads, arms, or body repetitively, with variations introduced subtly over a series of repetitions. Experimental studies indicate that this is optimal for attracting and holding a baby's attention [95].

By 4 months of age, infants will imitate the goals or ends of actions they observe, even doing so by different means [41, p.169]. Such behavior is evident throughout childhood and adolescence in such thoughts as "If she is doing that perhaps I should also" or "If he wants that, so do I." Indeed, it seems to be present in adulthood as well, even if it is embedded in such thoughts as "If she wants it, then maybe there is something good or useful about it that she knows and I don't." Related to the imitation of ends is the imitation of value or affective response. If, for example, an infant and mother establish common attention toward a robot, and the mother expresses joy and interest in the robot, the infant will be likely to touch it for an extended time and even pat or kiss it. If instead the mother expresses fear, most babies will not touch it at all and others will swat at it [96]. Such social referencing for valuation is also prominent in later childhood and adolescence, and continues throughout life, again probably with both reasoned and imitative components.

It is not possible to assign precise relative values to the reasoned and imitative components in adults when they follow the examples of others in what they strive for, desire, and like or dislike, but wide swings in style and fads and the technique and impact of advertising indicate that the role of imitation per se remains large. Moreover, the utilitarian value of imitation in infancy and childhood should not suggest that imitation be subordinated to hypothetical drives or other bases for goal-directed behavior in efforts to understand human nature and development. Imitation is a primary developmental process, and is evident when children imitate animals during play as well as when they imitate and acquire silly idiosyncrasies of those near to them, both of which are without extrinsic value. Indeed, as cross-cultural and subcultural analyses show, much of what is experienced as goals is the product of social processes. And as this book argues in general, even much of the reasoning offered as a rational basis for following the goals or valuations of others is itself imitative in its nature or origins. This is not to argue that there are no drives, desires, or needs intrinsic to human nature. It is to say that the need to eat, the need to procreate, and the need for physical self-defense are no more basic to human nature than the need for sensory stimulation, the need for affiliation, and the inclination to imitate. While the former are essential for life itself, it is the latter that give human life its particular form, that are the basis for human society and culture and for the dominance of the human species.

Language itself is only realized through imitation. While the capacity for language is linked to genetically encoded physical features unique to the human brain, and human infants babble even in phonemes specific to languages they have never heard, the language they do learn is the one they hear and imitate. They soon lose the capacity to utter sounds they do not hear and imitate, and, as the sad cases of children raised in social isolation attest, when there is little or no chance to hear and imitate speech, the capacity to speak and to understand language does not develop. New research on parental modeling and infant imitation of speech indicates that when talking to their infants, parents speak a "babyese." Certain key phonetic distinctions are exaggerated and extended in time [97]. Once the infant's developing auditory system can reliably make the perceptual distinction, parents gradually normalize their speech, again keeping the infant operating in its zone of proximal development. This seemingly automatic and universal parental behavior, so ideally suited for the infant's neurodevelopmental environmental needs, attests once again to the natural fit and importance of the parent-child dyad in shaping infant development. While some features of parental input are generally uniform across individuals, others vary as a result of either genetic or social environmental factors and create for each infant a unique mix of formative stimulation.

Acquisition of language provides a particularly informative example for understanding two aspects of the relationship between development-shaping social stimulation and characteristics of the brain itself. First, an important characteristic of imitative development of language is that the capacity is extremely strong for approximately the first 10 years of life and diminishes rapidly thereafter. Most children learn to speak and understand languages with ease during this period, but after that, acquisition of a new language is a laborious process that involves a very different sort of effortful study, even for teenagers. Here again, we see age dependence in the plasticity of neural structures. Imitative shaping in general probably follows a developmental time course similar to that so readily evident in acquisition of language. That is, after reaching puberty, imitative shaping of the brain to features of the environment is decreased, and as a result the individual becomes more set in his ways; a sense of individual identity begins to crystallize and the relationship between the individual and his environment begins to change. Increasingly, energy and effort are spent acting on the environment, or positioning oneself within the environment so as to maintain the symmetry and parallels between inner structure and outer reality that first developed as a result of the almost unlimited internal plasticity.

The second aspect of language acquisition that is of particular theoretical significance is the fact that if the usual language areas of the brain are destroyed in early childhood, children can still converse, read, and write. In most individuals the language function critically depends on the activity of similarly (but not identically) located regions within the left cerebral hemisphere. Indeed, these areas have a distinctive cytoarchitecture that distinguishes them from homologous areas in the right cerebral hemisphere and from similar areas in the brains of nonhuman primates. However, if illness or injury destroys these areas during the first years of life, language develops by relying instead on right-hemisphere structures [e.g., 98–100]. The language function is not entirely normal in these individuals, but the deficits are relatively subtle [101]. This has profound implications for understanding the relationship between the existence of special areas of the human brain that have distinctive anatomical features and special human functional capabilities.

No other animal has a language capacity that approaches that of human beings who have had their "language hemispheres" removed in early childhood. It is widely thought that the language capability in humans is due to special features of the brain that make human language possible. These have been linked to handedness and to functional specialization of the cerebral hemispheres, both of which are also much more marked in humans than in other primates. Consistent with this, lesions in these left-hemisphere areas in adults cause lasting deficits in language. One interpretation of these observations has been that the evolutionary development of these areas made language possible in human beings. Presumably, the thinking goes, these areas served some protolanguage function in now-extinct ancestors of *Homo sapiens* and evolved to the point of producing language through the combination of further mutation, selective mating, and environmental selection. What then do we make of the fact that if the very areas resulting from this critical evolutionary development are eliminated early in life, language develops nonetheless? For one thing, this observation is further testimony to the power of the human brain, in large part through imitation, to shape itself according to dominant features of the environment in which it develops. However, is it not also evidence of a dissociation between the capacity for language and the unique evolutionary processes and anatomical structures on which it is thought to depend? And does this dissociation not challenge the original understanding of the origins of the language capacity in humans? For if it is possible for humans to develop language without these structures, surely a theory that bases the evolutionary development of language primarily on the development of these structures is incomplete.

What it suggests instead is an expanded appreciation of the role of imitation, and social shaping of neurocognitive development, in the cumulative, transgenerational evolution of human beings. In this view, the evolutionary appearance of brain regions with some new and special functional characteristics enabled the development of language, providing a behavioral seed that grew into spoken language through generations of social cultural development, and then into written language through a more recent and more transparently social process. The relationship of the original behavioral seed to language as we know it is unclear, as is the initial, critical, generative functional change in the brain. Under most circumstances the language function continues to depend largely on the function of the special brain regions that provided the original behavioral seed. However, if these parts of the brain are destroyed, language can be subserved by different brain regions altogether. Once language exists in the interpersonal, cultural environment, the more general power and tendency of the human brain through imitation and related mechanisms to shape its growth around recurring features of the environment in which it finds itself allows the capability for language to develop in the absence of the areas that many thousands of years ago provided the functional characteristics that allowed human beings and human societies to first develop languages. This capability is subserved instead by more ordinary brain tissue. Moreover, as indicated, in the few known and tragic instances when children with physically intact brains were raised in near-total social isolation, the children had little or no language ability. One must conclude that language is not a property of the human brain but rather of human society and culture. If all living people were rendered permanently speechless and illiterate, their offspring and succeeding generations would be unable to speak despite having normal brains, and language, this most distinctive of all human characteristics, would be lost to the human species.

Internalization and Identification

Of all the objects in the infant and child's sensory experience, animate or inanimate, the most prominent are the parents and/or a select few other primary caregivers, followed in varying degree by siblings. These figures are imitated piecemeal: the ways they direct their attention, feel about things, think about things, organize things in the world and in themselves, think about themselves and their activities, and do this or that. It is they also who instrumentally lend their organizing schema, and other frontal lobe functions, to incorporate the developing child's newly acquired motor and perceptual skills into larger behavioral units, and in doing so shape the development of the child's own schema, behavioral routines, and frontal lobe functions. The composite effect of these experiences is to create in the child a mix of qualities that resemble those of these prominent figures. We call these composite effects internalization or identification.

Psychoanalysis has provided important insights into these processes. These insights come from a different type of data and different type of data acquisition process than we have considered so far. Instead of direct observation of developing children in their homes or in a laboratory, psychoanalysts listen to adults describe their thoughts, feelings, and experiences in the world; observe their behavior during analytic sessions; and study the structure and patterns of their speech. In addition, and probably most important, the psychoanalytic situation promotes the creation of dyadic patient-therapist systems that are operative within the analytic sessions. The analyst observes the roles and characteristics assumed and exhibited by the patient in the dyad and the feelings the patient experiences in those roles. Moreover, the analysts monitor the roles and feelings they themselves experience as the complimentary half of the dyad. Many of the observations made in these ways by psychoanalysts, and the accompanying theoretical deductions, are remarkably similar to those already presented and derived from other disciplines. This concordance between observations made from very different theoretical and methodological vantage points adds credence to both sets.

In the parlance of psychoanalysis, the ego is the part of the psyche that perceives, reasons, thinks rationally, and acts. Writing in 1926, Fenichel stated that changes in the ego as a result of the adoption of the characteristics of another person have long been familiar to psychoanalysis. Called identifications, they were presented as an alternative relationship to that of a libidinal involvement with an important person (or object) in the external world [102]. Freud described identification as "the assimilation of one ego to another one, as a result of which the first ego behaves like the second in certain respects, imitates it and in a sense takes it up into itself" [103, p. 47]. Greenson says that in identifying with a person (or object) the self is transformed, becomes similar to that person, and adopts the person's behavior, attitudes, feelings, posture, etc. [104]. Reich, in language similar to that used earlier in this chapter, explains that "the child simply imitates whatever attracts his attention momentarily in the object. . . . Normally these passing identifications develop slowly into permanent ones, into real assimilation of the object's qualities" [105, p. 180].

Much of psychoanalytic thinking rests traditionally on the starting assumption or view of individuals as independent agents or entities, with a central ego, I or self. The problem from this starting point is then having to connect individuals with one another, a problem solved in psychoanalytic thinking primarily by the posited role of drives that lead or push one individual to act upon another. In the course of such actions, interpersonal contact is made and relationships formed, but the nature of these contacts and relationships is influenced fundamentally by the drives that initiate them. Fenichel sees in identification a quite different type of relationship between individuals and people or objects in their environment [102]. A critical difference between these two types of relationships is the

degree of separateness among individuals assumed as the starting point. Fenichel comments further on this difference, remark-
ing that in identification the barrier between the ego and the
world of things is abolished but without an impairment of
reality testing. The ego has taken over the form, the interests,
and the function of the object [102].

More recent psychoanalytic writers have expressed positions
even more similar to the one being developed in this book.
Describing Winnicott's ideas, Ogden explains that "a new
psychological entity is created by mother and infant that is not
the outcome of . . . simple summation of parts. . . . It is . . . the
mother-infant that is the unit of psychological development"
[106, p. 171]. Winnicott himself writes that the behavior of the
environment is part of the individual's own personal development [107] and Greenson states that at early stages of development perception implies transformation of the self [77]. So,
while the theory developed in this book places much less
emphasis than does classic psychoanalytic thinking on drives
such as love, aggression, and hunger in shaping human development and relationships, some psychoanalytic thinkers have
come to a position similar to the present one. Moreover, despite
holding some very different tenets regarding the degree of separateness and the nature of relatedness, psychoanalytic observations of identification provide additional evidence of imitative
shaping of the developing human mind.

Impressive similarities between psychoanalytic observations
and those of psychologists such as Vygotsky and Luria are also
noteworthy, especially given the differences between the two
in theoretical assumptions, cultural context, and observational
techniques, and the independent development of each school
of thought. In *An Outline of Psychoanalysis*, Freud wrote, "A

portion of the external world has, at least partially, been abandoned as an object and has instead, by identification, been taken into the ego and thus become an integral part of the internal world. This new psychical agency continues to carry on the functions which had hitherto been performed by the people in the external world" [108, p. 205]. Here again the contrast between libidinal relationships with an object (or person) and identification is evident. Here also is explicit recognition that functions performed by other individuals in the child's life become part of the child's competence by identification. Similarly, Hartmann, a psychoanalytic colleague of Freud's, wrote that introjection is an ontogenetic process that allows internal regulation to replace environmental regulation and internal processes to replace external action [109]. Writing more recently, Loewald states that the psyche now is seen as an emerging organization that evolves through active and increasingly complex interchanges with people. These interactions and relationships become internalized, creating a system within the individual of interactions, relationships, and connections among different elements to produce the intrapsychic structure [110]. These views from psychoanalysts are remarkably similar to Vygotsky's description quoted on pages 112–113, of how a "function which hitherto was shared between two people now becomes a method of internal organization of the psychological process" [74].

An important feature of these processes, then, is that what becomes internalized are not only the qualities of other, more mature individuals, but also interpersonal or even multiperson processes that had not previously existed in any particular individual. That is, the qualities of the developing individual arise from interactive combinations of processes based on several individuals, just as, by analogy, an individual's genetic makeup is a unique combination of parental components that can produce qualities that do not exist in either parent. The creation of interindividual variation in functional characteristics is enhanced through internalizations of interpersonal rather than intrapersonal processes, just as genetic variation is enhanced through sexual rather than asexual reproduction.

Psychoanalysts have expanded the notion of internalization to include identifications with other animals and with human artifacts. As Schafer writes "the subject may . . . identify with representations of non-human creatures and things as well as those of other persons. Identification with pets, wild animals, and machines, to give a few examples, are not rare. He may also identify with representations of non-living persons, such as fictional characters, ancestors and significant figures of local, national or world history or myth" [111, p. 307]. Powerful examples of identification with nonhuman creatures are evident in the traditions of certain Native American peoples. Young men, for example, were named for animals with whom a special characterological connection was assumed, and the passage into manhood sometimes included spending several days alone in the forest meeting the spirit, and perhaps a living example, of the animal and thereby assuming important qualities of that animal.

The array of fictional, mythical, and inanimate objects available for imitation and identification increases from generation to generation, as the machines, fictional characters, and heroes of the previous generation become available to the next, and as new forms of information technology make representations of objects more widely available. Interest in many of these "secondary" objects can reflect already-formed characteristics of an individual, rather than relationships that transform the individual, as will be discussed shortly. However, psychoanalysts suggest that interactions with these objects can be formative as well.

With the wide array of people, relationships, and other objects with which a child may identify, one might predict that the integration, harmony, and internal consistency of the resulting internalizations would not necessarily be a matter of course. And indeed, psychoanalytic observations indicate that conflicts among internalized qualities are not uncommon. Freud stated specifically that conflicts among different identifications are not purely pathological [112], and Erickson considered a major task of adolescence to be the subordination of childhood identifications to produce a lasting pattern of inner identity [113]. Here again, social influences are profound, according to Erickson, as adolescents struggle to integrate their personal mix of identifications into a stable personal identity that also fits with social mores and available social roles. When doing this, the individual chooses an ideology, from among those presented by his or her society, that offers a seeming correspondence between inner and outer worlds and that provides "a geographic-historical framework for the young individual's budding identity" [113, p. 284]. At the same time, the adolescent strives to find a niche in society that "will reconcile his conception of himself and his community's recognition of him" [113, p. 268].

Chapter 5 will address in detail the force and persistence with which individuals will work or fight to maintain this niche and their ideology. It is interesting at this point, however, to note the developmental changes in relationships between object and perceiver suggested by the oral, anal, and phallic developmental stages that figure prominently in psychoanalytic thought. In the oral stage, the highly plastic and rapidly developing infant is easily transformed by multiple identifications and internalizations. By the anal stage, the child's psyche has a structure of its own; identifications are more grudgingly fitted in and they lead to more limited changes in the child. In the phallic stage, the child begins to act on, alter, and shape his or her environment instead of being shaped by it. From the perspective of individual subjective experience, at least as constructed by contemporary Western society, the phallic stage marks the beginning of the creative and instrumental action in the world that has made humankind the master of its environment. However, much or most of this action is aimed at maintaining a fit between an increasingly fixed inner world and a continually changing external world.

In another paradoxical disjunction between individual and group phenomena, it is really the early shaping of the infant and child's psyche by the human-influenced environment, with the unique mixing of qualities from different adults and the internalization of historically influenced interpersonal processes, that has led to new generations with new qualities and to the rapid evolution of human capabilities, human culture, and human dominance. While the action of developed individuals expresses and instantiates this intergenerational plasticity, it is fundamentally conservative within the individual, seeking to mold and keep the external world consistent with the individual's established inner structures.

Psychoanalysts have identified interesting examples of this conservatism, which is called projection and projective identification. In projection, parts of the self are projected out onto the world, and the world is perceived and/or experienced as if it had these properties. Inner-outer consistency is maintained by seeing the outer as if it were identical with the inner, even if it is not the case—a far cry from the early imitation of the outer and resulting transformation of the inner. As the individual relies on projection, he or she is blind to novelty. Cognitive psychologists refer to similar processes when describing the role of internal, experience-based schema in focusing attention and in organizing and interpreting perceptions. Such schemas, for example, assign different natures to the yellow liquid in a plastic cup held by a standing man dressed only in shorts, depending on whether he is by a swimming pool or in a doctor's office.

In projective identification, an individual projects part of the self onto another person, and then identifies with that other person. To maintain the fit, the person upon whom the attributes are projected is pressured by subtle and not-so-subtle means to act in a manner consistent with the projected qualities. The psychoanalytic situation is structured to foster projective identifications. By monitoring the ways in which he or she is perceived and the pressures he or she experiences to feel or act a certain way, the analyst learns about the internal psychic structure of the patient.

The following clinical vignette provides a dramatic example of this process [114]. The patient was a 40-year-old businessman who described his father as a powerful, irate, extremely demanding, and sadistic man who punished his children severely for minor misbehaviors. He described his mother as shy, socially withdrawn, and completely submissive to his father. The father's qualities, the mother's qualities, the patient's relationship to his father and to his mother, and the relationship between the patient's parents had all been internalized in the patient and affected the ways he perceived, understood, and acted in the world. Shortly before the treatment session of interest, the patient had begun dating a woman who worked in the large hospital complex with which the therapist was associated. The evening before the session, the patient had asked his lady friend whether she knew his therapist and, upon finding out that she did, pressed her for information about him. The woman then asked the patient, in what he considered to be an ironic manner, whether he was a patient of the therapist. The patient was convinced that the woman had known this all along. When he confronted her with what he considered to be this fact, she apparently became concerned and suggested they "cool" their relationship.

The patient began the therapy session the following day by stating that he felt like punching the therapist in the face, and demanded an explanation. When the therapist asked "for what?", the patient accused him of "playing innocent" and became further enraged. The patient went on to describe the events of the previous evening and accused the therapist of having called the woman, of having told her about all the patient's problems, of warning her against the patient, and of causing the end of the relationship. When the therapist suggested that these ideas were related to a previously expressed fantasy that he, the therapist, owned all the women of the institution and was a jealous guardian of his exclusive rights to them, the patient became further enraged and accused the therapist of misusing interpretations to deny the facts. The therapist became frightened by the patient's mounting rage and said that he did not feel free to talk openly because he was not sure the patient would not act on his feelings. The patient then asked the therapist whether he was afraid of him and the therapist said he was. The patient smiled, continued discussing the woman in an angry but no longer threatening manner, and soon said he no longer thought the therapist was lying. He added that all of a sudden the whole issue seemed less important to him, and that it felt good that the therapist had been afraid of him and had acknowledged as much. The therapist realized that he himself also now felt that the whole relationship with the woman was less important. After reflection, he said that a fundamental aspect of the relationship of the patient to his father had just taken place, namely, the enactment of the relationship between the patient's sadistic father and the patient himself, in which the therapist had taken on the role of the frightened, paralyzed child and the patient the role of his rageful father who secretly enjoyed intimidating his son.

The patient had projected the frightened, childlike and impotent component of himself onto the therapist, forced the therapist to feel and act in a manner consistent with this projection, and played out an internal struggle by assuming the role of the powerful, angry father. This was turned to therapeutic advantage when the therapist acknowledged his fear of the patient and thereby decreased the patient's own sense of humiliation and shame at being terrorized by his father. That it was safe to express rage at the therapist without destroying him made it possible for the patient to tolerate his own identification with his enraged and cruel father. The patient then said that perhaps he had frightened the woman because of his inquisitorial style in asking about the therapist and that his suspiciousness about her attitude toward him when she acknowledged that she knew the therapist might have contributed to driving her away.

Play

While interactions with neurocognitively mature adult caregivers have more significant effects on the child's development, interactions with peers, especially in the form of play, are also important. Play is seen little among reptiles, and in mammals it is most common and most elaborate in primates [115]. It is considered a puzzling anomaly among behaviors of animals, since it consumes large amounts of energy yet almost by definition seems to have no instrumental goal [116]. At the same time, its ubiquity among mammals and the energy devoted to it suggest biological importance. Indeed, if rats are deprived of the opportunity to play by being socially isolated for 8–24 hours, there is a "rebound" increase in the time spent playing once they are reunited with their peers, speaking again to some biological significance [117]. Kittens weaned artificially early play more than those weaned later [118,119], and those fed with a stomach tube rather than being nursed begin playing at an abnormally young age [120]. Both observations support a role for play in the social developmental processes discussed earlier. A number of purposes or functions have been proposed for play in animals, including development of the peer group relationships, social skills, and dominance hierarchies important for adult community life, and development through practice of motor skills necessary for hunting and self-protection.

Social isolation of rat pups has its greatest impact on development when it occurs between the ages of 20 and 45 days, a period coinciding with the time that rat pups most actively engage in rough-and-tumble play. This isolation and associated play deprivation significantly affects cognitive development. For example, such rats are slower to habituate to new settings [121] and slower to reverse previously learned discriminations [122,123]. One hour per day of rough-and-tumble play protects against the effects of isolation, while daily contact with a drugged, nonplaying peer does not [124]. Thus, play appears to affect cognitive development, even in rats and even when the play is primarily motoric. The role of play in human cognitive development may be greater because it lasts for years rather than days, is highly varied in nature, and includes activities that are primarily cognitive and essentially social. Here then is another avenue for social and cultural influence on important aspects of brain development.

When play is viewed as a means to practice and thereby facilitate the development of the adult behavioral routines essential for survival, such as hunting and self-defense, it is often assumed that the behaviors are genetically predetermined but need exercise to develop. Although the behaviors are more complex and the requisite practice correspondingly greater, the process is seen as essentially similar to what appears to be involved in developing the ability to walk and run [125]. The ability is innate and instruction is not necessary, but competence is not full born and initial feeble attempts are necessary to develop the required strength and coordination. While this is probably one function of play, the nature and role of play appear to vary from species to species [125]. Given the evidence already cited of the sensitivity of primate brain development to sensory stimulation and motor activity, it is likely that an activity as rich in stimulation and action as is play, and indeed an activity particularly prominent in primates, also affords the opportunity for postnatal shaping of behavior and related brain structures. What is of interest from this perspective is the possibility that play offers an opportunity for social interaction to affect the ultimate characteristics of adult behavior rather than simply to facilitate its unfolding.

For play-related social interaction to alter the course of behavioral development, there must be some source of change in the nature of the play itself, for if play remains the same generation after generation, it will only lead to the same adult behaviors. Changes in the physical environment could produce such changes in the play of animals, leading to alteration in adult behaviors, although I am unaware of any studies demonstrating such effects. In human beings, however, adults actively intervene, direct, and proscribe the play of children; the objects with which children play are for the most part made by human beings; and social and cultural processes transform the nature of play. Indeed, in biological terms, the whole of formal education is perhaps most appropriately seen as a human extension of play.

Human play is also remarkable for the extent to which it is cognitive rather than motoric. This is readily evident in chess, card, board, and computer games. However, it is also true in fundamental aspects of physical and sports games, as pointed out by Fagen and Vygotsky, who approached the same problem and arrived at similar conclusions from different starting points. Fagen [126], citing Millar [127], writes that an essential feature of play is that it occurs in a paradoxical context in which rules for interacting with the environment are treated as objects of discourse rather than as plans controlling behavior.

Vygotsky argues that imagination and rules are the essential features of human play:

Every imaginary situation contains rules in a concealed form, [and] every game with rules contains an imaginary situation in a concealed form. The development from games with an overt imaginary situation and covert rules to games with overt rules and a covert imaginary situation outlines the evolution of children's play.... it is here that the child learns to act in a cognitive, rather than an externally visual realm by relying on internal tendencies and motives and not on incentives supplied by external things. ... in play things lose their determining force. The child sees one thing but acts differently in relation to what he sees. Thus a condition is reached in which the child begins to act independently of what he sees. [74, pp. 95–97]

As the hold of the external physical environment on the attention and actions of children is loosened, culture fills the gap. Through hours and hours, weeks and weeks, and years and years of play extending into adulthood, if not throughout the life-span, and often involving repetition of the same or highly similar activities, society provides rules that affect development and that create lasting neurocognitive structures.

Summary

A key feature of the human brain, as discussed in chapter 1, is that its functional capabilities result from the integrated activity of large numbers of richly interconnected neurons. Chapter 2 showed that the extent and particular patterns of these interconnections depend upon sensory stimulation, and that humans and other mammals seek this stimulation. Sensory stimulation can serve as a reward in conditioning experiments, as does food; sensory deprivation in infants limits brain development; and sensory deprivation in adults compromises function and increases stimulus-seeking behavior.

A second key feature of the mammalian brain, again as discussed in chapter 1, is the evolutionary appearance of limbic structures essential for social and familial behaviors. This chapter has reviewed evidence that sensory stimulation resulting from familial and other social interactions has even more profound effects on the development of human brain function and is responsible for advances in human capabilities at a rate that far exceeds that possible through the evolutionary processes that led to the initial appearance of the human species.

Interactions between biobehaviorally mature adults and highly plastic, unstructured infants and young children shape the development of the infants and children. These interactions are in part directly physical (body-to-body), especially in the first months of the infant's life. More important, through instrumental parenting, adults create the physical world experienced and acted in by children, and interactively intervene in the object-directed activities of children. In these interventions adults provide frontal lobe functions of which the child is not yet capable (e.g., memory, planning, organization, and strategy) and that enable the earlier maturing sensory, motor, and association areas of the brain to develop as components of more complex functional systems. The frontal lobes, which are so much larger in humans than in other primates, then develop under the influence of activity from the externally structured, earlier maturing, cortical regions as well as from continuing social interactions. Language, itself a social and cultural development of brain capability, has radically increased the power of adults to influence the development of the brain in children, both through interpersonal interactions and through culture. Imitation and internalization are more automatic, equally important processes through which infants and children shape themselves after the adults they observe and with whom they interact.

There is a striking convergence in the descriptions of these processes by researchers from different theoretical and investigative traditions, and in the conclusion that through these processes, what was first external and interpersonal becomes internal structure. Adolescence and young adulthood are occupied with the dual tasks of integrating internal structures derived from multiple sources into a functionally coherent whole, and articulating a personal ideology that leads to a niche in the general social matrix that is consistent with the internal structures.

These progressive developments in children are expressed or instantiated by their actions in the world as adults. These actions, then, alter the rearing environment of the next generation and complete the cycle that has led to such rapid evolution of the functional characteristics of the human brain. It is important to note, however, that while the actions of adults in altering the environment (including culture) are an essential part of the intergenerational, social process that so effectively promotes change and development of human capabilities, their motivation from the individual's perspective is fundamentally conservative. That is, by the time an individual is able to act on the world, his or her efforts are aimed primarily at expressing an already formed inner world. This may involve creating new structures, activities, or organizations in the external world that reflect their inner worlds. It may also involve efforts to prevent changes in aspects of the external world that fit with their inner worlds. These latter efforts are the focus of the concluding chapters.

II The Neurobiology of Ideology

Ideology: a systematic scheme or coordinated body of ideas or concepts especially about human life or culture

Webster's Third New International Dictionary

4 Self-Preservation and the Difficulty of Change in Adulthood

Such tricks hath strong imagination,
That, if it would but apprehend some joy,
It comprehends some bringer of that joy;
Or in the night, imagining some fear,
How easy is a bush supposed a bear!

—William Shakespeare, *A Midsummer Night's Dream*

Exile is strangely compelling to think about, but terrible to experience. It is the unhealable rift forced between a human being and a native place, between the self and its true home: its essential sadness can never be surmounted.

—Edward Said, "Reflections on Exile," in *Out there: Marginalization and Contemporary Cultures*

Thus far, the argument has been that the neural basis of most human behavior, and certainly of most capabilities that distinguish human beings from other animals, is to be found in the integrated action of cell assemblies and not in the properties of individual neurons. The interconnections among neurons that create these cell assemblies can only develop with adequate sensory stimulation, and the nature of the interconnections depends upon the nature of the stimulation. Human beings seek out the sensory stimulation necessary for their growth and development, and even adults function poorly when deprived of stimulation. Particularly important stimulation comes in the form and actions of other human beings. There are both differences between and commonalities among the stimulations received by different individuals. The differences contribute to variation even among siblings in patterns of neuronal interconnections and cognitive and personality characteristics. The commonalities in experience lead age cohort groups within a given culture to have common characteristics. Of particular significance from a biological perspective is the fact that these common experiences result from relatively enduring changes made in the environment by other human beings. No other species makes changes in the environment experienced by their offspring to an extent that approximates those made by human beings. This feature of human beings has played an important role in the rapid evolution of human capabilities and dominion. The fact that the human brain shapes itself to a progressively changing and increasingly human-made environment has been of as great or greater importance than innate abilities in the human mastery of the animate and inanimate environment. Indeed the evolutionary gain in neural plasticity that distinguishes human beings seems to have been at the cost of innate capability. Based on the behavior of a few unfortunate individuals raised with adequate physical nurturance, but with limited sensory and essentially no sensorisocial stimulation, a group of such individuals would be unlikely to fare as well as our nonhuman primate cousins.

Chapter 3 focused on the role of interpersonal relationships in development of the brain. In this context the processes of projection and projective identification were discussed. The emergence of these processes is an early aspect of the fundamental reversal of the relationship between the individual and the environment that takes place largely during adolescence and early adulthood. Until then, as just outlined, the highly malleable inner world is shaped by the external world. For the remainder of life, the individual largely acts to alter the external world to match an increasingly inflexible inner world. Projective processes have two components. The first alters the perception and experience of the external world according to preexisting internal structures. The second alters the course of events in the external, in this case primarily interpersonal, world in such a way as to increase the likelihood that subsequent events will be consistent with the preexisting internal structures. In the case of projection, the individual acts toward the external world as if it exists in the form of the internal structures, and this action affects the way the world responds, often creating a response that is consonant with the projection. In the case of projective identification, the individual more actively manipulates other individuals to act in a manner consonant with the individual's own internal structures.

Although these efforts to manipulate the environment are limited and primitive compared with others discussed later, they signal a change in the tide. This outgoing tide can be fraught with danger and destruction in ways discussed in the last part of this chapter and in the next. Suffice it at this point to identify a central feature of individual human development: Learning and action are in an inverse relationship throughout the lifespan. We learn the most when we are unable to act. By the time we are able to act on the world, our ability to learn has dramatically diminished.

This chapter begins with presentation of further evidence that, once established, internal structures shape the perception and experience of the external world to their own form. This demonstrates that higher-order cognitive structures, and social-emotional structures, are self-perpetuating in ways similar to the perceptual structures of the visual cortex described in chapter 2. The second section provides evidence that this process is not simply a necessary consequence of having developed structured perceptual and thought processes, but also derives from the fact that a consonance between internal and external structure is experienced as pleasurable, while a dissonance is an unpleasant source of psychophysiological tension. The third section considers what happens when forces beyond a person's control leave them in an environment changed to a degree that exceeds the ability of selective perception and experience to maintain a match between internal structures and the external world. One such situation is the death of a spouse; here a major component of the individual's external and internal social world has been removed from the external world. Another such situation is immigration into a new culture; here the human-made cultural environment, including language, food, customs and more, and often the physical environment, change dramatically from the environment that had shaped the individual's internal structures. The subjective discomfort and objective dysfunction that regularly result from these situations provide more evidence of the importance of internal-external consonance in human well-being. In both instances, the extended and arduous effort required to restructure the internal world so that it matches the new external world attests to the difficulty and importance of reestablishing this consonance.

Further Evidence that Internal Structures Shape the Experience of the External World to Their Own Forms

It is common knowledge that different individuals see the same thing in different ways. Each selects, emphasizes, and values different components of an experience. When a continuous stream of stimulation is broken into discrete events, it is not uncommon for two individuals watching the same series of events to disagree as to whether some particular event even occurred. What would café life be like without such interindividual differences? Indeed, most people have a certain fascination with these differences and an interest in articulating and arguing their own views. Such discussions, however, can get quite heated, at times even leading to physical altercations, providing a glimpse of how important it can be to an individual that their view be confirmed or at least allowed to remain intact. For the most part, parties to these discussions maintain their respective positions throughout. However, it is an often-exercised prerogative of social power to have one's view of things adopted by subordinates at the expense of their own views. To avoid such situations, most people aggregate into affiliative networks with other like-minded individuals. An important aspect of many social transactions among adults is mutual confirmation of perceptions and valuations of events that have been questioned by contradictory articulations from other sources.

The way an individual sees and experiences an event can often be predicted from known characteristics of the individual. That is, individual differences in perceptions and experiences of events are not capricious, or are not usually the result primarily of some event that by chance preceded the one in question, but rather result from stable, experience-based, internal structures.

Two commonplace examples make this clear. One is allegiance to sports teams. A Yankee fan and a Red Sox fan will see the relative merits of players on the two teams, and the "facts" of specific plays during a game between the teams, in very different and highly predictable ways. Such allegiances are most strongly established during childhood and are often fostered by family and peer group attitudes and shared experiences, but become enduring characteristics that influence perception, experience, and interpretation of external events. Some individuals even feel good when "their" team is winning and bad when it is losing, demonstrating the impact of fluctuations in external events on fluctuations of internal state once the connections between internal and external have been forged, even in an arena with no instrumental significance to the individual. Connections can be so strong between sports fans and the teams with which they identify that they feel more confident about their own competence to perform mental and physical tests, or attract members of the opposite sex, after their team wins than after it loses [1]. They are more likely to wear clothing with their team's logo the day after their team wins than the day after it loses, and they claim a share in their team's successes by saying "we won yesterday", but distance themselves from their team's failures by saying "they lost yesterday" [2,3]. The physiological correlates of these subjective and behavioral responses are remarkable. Testosterone levels in male Brazilian soccer fans rose on average 28% after their team defeated Italy in the 1994 World Cup, while levels in their Italian counterparts dropped 27% [4]. Physiological arousal, measured by changes in electroencephalograms, heart rate, and perspiration, is marked when fans are shown pictures of the stars from their team making game-winning plays, but absent when they are shown pictures of other athletes making comparable plays for other teams [5]. And hormonal and other physiological changes in male hockey fans are similar to those in the players they are watching.

A second, not totally dissimilar example is the impact of political party membership on the perception and valuation of events. No scientist could have designed a better demonstration of these effects than that provided by President Clinton's relationship with Monica Lewinsky and the ensuing congressional deliberations. Most agreed that the events in question were adequately captured in recordings of relevant phone calls, legal depositions, and legal testimonies. Indeed, for the deliberations in both the House of Representatives and the Senate there was a generally accepted body of evidence from which to draw inferences and make decisions. Thus, from the point of view of a scientific experiment, there was a common, reproducible stimulus provided to all subjects. Similarly, the relevant independent variable or subject characteristic, membership in the Democratic or Republican party, was unambiguously defined. Moreover, the subject pool was largely drawn from intelligent and well-educated sectors of the American population. The results were dramatic. Almost to a person, members of Congress selected component parts of the evidence, valued those components, and reached conclusions as to whether certain things had or had not happened in a manner consonant with their party membership. They each did this while stating that they were doing their utmost to exercise their individual judgment before the view of millions of Americans and so as to withstand the test of history. Moreover, their decisions were of little instrumental value to them individually or even to their parties collectively, as the Democrats voted to keep in office a president who was a political liability and the Republicans voted to put into the presidency a man who would then run for election as a sitting president. Similar processes were evident in the population as a whole, as polls indicated that most Democrats were against impeachment and conviction while most Republicans were in favor of these measures. Political party membership is highly familial, but not genetically transmitted. A person comes to see her or himself as a member of a political party because adults significant in that person's upbringing were members of that party. Along with party identity comes a set of beliefs about a range of social issues. These beliefs influence the perception, experience, and valuation of many things, and the actions based on them. People are even more likely to marry someone from the same political party than from another, even though they mix with people from all parties in school and social settings.

Documentation and analysis of the ways in which internal psychic structures influence perception and valuation of external stimuli was a major occupation of experimental psychology for several decades of the past century. Two types of processes have been described. One set characterizes the way an individual perceives and remembers, independent of the content of the experience. The other processes are content based and influence meaning, salience, organization, and valuation of perceptions. The first set includes stable individual differences in the extent to which people scan across or focus on a part of the sensory field, extract information from its embedding contexts, ignore task-irrelevant stimuli, minimize or maximize differences between new and old precepts, use fine-grained or coarse categorizations, adhere to or suspend a strict reality orientation, augment or diminish responses to novelty, or generally expect a positive or negative outcome [6–8]. It is not known to what degree these characteristics are influenced by genetic makeup, but they are thought to be substantially influenced by interpersonal and cultural factors during childhood development [e.g., 6,9,10].

When adults in Japan and America are shown an animated underwater scene in which one large fish swims among smaller fish and other marine life, Japanese subjects usually describe what they have seen beginning with a general description of the scene and include multiple statements about the relationships among objects, while Americans are more likely to begin with a description of the big fish and make only half as many comments about the relationships among objects [11,12]. When shown the same big fish in a second underwater scene, the Americans are more likely than the Japanese to recognize that the big fish is the same. When adults in Korea and America were asked to read essays and were told that the authors had been instructed to espouse the particular views expressed, both groups still expressed beliefs that the writers actually held the views presented in their essays. After the subjects were themselves required to write essays according to instructions, the Koreans, but not the Americans, decreased their estimates of how strongly the original essay writers believed what they wrote.

When adults in China and America were asked to analyze conflicts between mothers and daughters, the Americans were more likely than the Chinese to support one side or the other, while the Chinese were more likely than the Americans to express merit in the position of both parties to the conflict. Similarly, American subjects express greater support for a particular position when weak arguments against the position are presented along with strong arguments in favor of it, than when the strong supporting arguments are presented alone. In contrast, East Asian subjects moderate their support for a position when the weak arguments in opposition are presented along with the strong arguments in support.

The effects of content- (and therefore experience-) based internal structures on perception and valuation have also been repeatedly demonstrated in a variety of experiments. Some of these effects are relatively short lasting and result from other recent perceptions or from fluctuations in internal physiological state. For example, in one experiment the subjects were asked to recall a list of twenty words. If the subjects were food deprived and shown the word "cheese" prior to presentation of the word list, they recalled more cheese-related words [13] from the list. If they were also asked to read and rate for literary style either a paragraph about the pleasures of eating pumpkin pie or a control passage, the subjects who read about pumpkin pie recalled still more cheese-related words [14,15]. In this experiment, then, the subjects' internal physiological and mental status altered their recall and probably some aspect of perception and coding of incoming precepts. For experimental purposes, receptive mental states were themselves manipulated by external cues, but the effects of these cues depended upon activation of internal, culture-specific and experience-dependent memory networks. Such a sequence is a constant occurrence of everyday life, where some stimulus activates internal structures that then influence subsequent perception.

Other experiments have demonstrated the effects of enduring internal characteristics on perception without the need to experimentally prime those structures. When subjects are shown incomplete or ambiguous stimuli, for example, they provide information from internal sources to complete the perception. While this may be a reasonable and efficient perceptual strategy, especially in a world with a continuous stimulus load that is well in excess of processing capacity, it is nonetheless a demonstration that people see what they expect to see. Those expectations are based on past experiences that have led to enduring expectations that are all the more difficult to alter as they come to shape subsequent experience to their own form.

In another experiment, the subjects' values were first assessed with a self-report rating scale, and they then were shown words more or less consonant with their personal values [16]. Words were first presented on the viewing screen for such a short time that the subjects could not read them (less than 10 msec). The exposure time was gradually increased so the threshold for legibility could be determined. The subjects were able to identify words that were more consonant with their personal values at lower exposure times, providing experimental evidence of the effects of personal value systems on simple perception. Another series of experiments has demonstrated similar effects of cultural value systems. For example, when 10-year-old children were asked to adjust the size of a circle of light to match the size of either a coin or of a cardboard disk the same size as the coin, the size estimates of the coins were consistently larger than the size estimates of the disks. Moreover, the relative discrepancy increased with the value of the coin [17,18].

Several theoretical concepts have been offered to organize these findings. Individual differences in perception have been seen to arise from part-cue response characteristics; each individual responds to a limited and different sample of the information input, with the selection governed by their personal makeup [19]. Perception has been seen as hypothesis driven, with the hypothesis representing a predisposition to organize environmental input in specific ways [20]. The stronger a hypothesis or belief system, the less input is needed for confirmation. More recently, past experience has been seen as organized in schema that provide the basis for the selection and interpretation of new stimulus input. Such schema arise both from individually specific and culturally shared experiences. A medical doctor seeing a picture of a man wearing nothing but shorts and holding a small cup of yellow liquid might immediately place the man in a medical office setting, while a swimming instructor seeing the same picture would think of a pool party. Cross-cultural differences in schema can be marked. The fact that Eskimos have names for many different types of snow provides an example. Through many generations of living in an environment where such distinctions are of practical use, Eskimos have developed the ability to recognize types of snow that have different implications and have passed this ability on to their offspring, along with a language to describe the types of snow. Individuals living in more temperate and more urban environments do not have similarly differentiated schema for types of snow and do not as readily perceive or encode these features of their winter environment.

In general, many of these experiments and theories refer to a form of internally driven prejudgment of what is to be perceived. Studies of prejudice, i.e., distorted and negative presuppositions about people of a different race or culture, provide still more examples of the general process. Here perceptual and value distortions about another person can be particularly strong despite little or no actual experience, even with the class of people of which the person is a member. In these instances the prejudicial beliefs derive directly from sociocultural input, including the internal structures of important adults to whom the individual is exposed during childhood.

Recent experiments demonstrate that these effects are deeply ingrained in affective structures. In these experiments individuals are shown words on a computer screen (targets) and asked either to simply read or in some way evaluate the words. If prior to presentation of the word to be categorized, another word or a picture is flashed on the screen for so short a time that it does not consciously register (prime), it can still affect the time it takes for the subject to read or evaluate the subsequent target word. If the briefly flashed prime has the same emotional valence as the subsequent target word, then response time to the subsequent word is usually shortened. If it is of the opposite valence, the response time is generally lengthened. This effect is so reliable that it can be used to determine the emotional valence people assign to either the primes or the targets. In this manner white Americans have been shown to experience pictures of unfamiliar black Americans as emotionally negative and pictures of other white Americans as emotionally positive [21–23]. In another similar experimental procedure, subjects are shown two stimuli at once on a computer screen. One might be a word with a pleasant or unpleasant meaning, and the other a picture of an insect or a flower [24]. With one hand they push a button to indicate whether the word was of the pleasant or unpleasant type. With the other hand they push a button to indicate whether the picture was of an insect or a flower. The responses are quicker when the two stimuli are of the same affective value (flower and pleasant word or insect and unpleasant word) than when they are of opposite affective value (flower and unpleasant word or insect and pleasant word). This effect is also so consistent that it can be used to determine the affective valence of stimuli by pairing them with stimuli of known valence. With this method, it has been shown that Korean-Americans experience Korean surnames as more emotionally positive than Japanese surnames, while the reverse is true for Japanese-Americans [24]. Such prejudice demonstrates the effect of interpersonal modeling and sociocultural education on internal processes that alter the perception and valuation of stimuli. White subjects who showed more negative automatic attitudes toward blacks on these laboratory tests also had lower ratings of friendliness in interactions with a black researcher and maintained less eye contact with a black than with a white interviewer, demonstrating the significance of these processes in real-world social behavior.

Brain imaging studies have shown that the degree of such social prejudice is associated with the degree to which the amygdala is more strongly activated by pictures of the faces of black than white people [25], demonstrating a link between these behaviors and a brain structure known to be important in social and emotional behavior. This association was not present when the subjects viewed faces of famous black people with whom they were familiar, and the generally greater activation of the amygdala seen with unfamiliar black compared with white faces was absent with familiar faces. Familiarity implies the existence of internal representations that correspond to the external stimuli. Here then is evidence that unfamiliar social stimuli for which there is no "close enough" corresponding internal representational structure are experienced as unpleasant. The converse of this, pleasure associated with the familiar, is discussed later, and the deep difficulty people have with unfamiliar types of people is the topic of the next chapter.

Thus, everyday observation and many experimental studies make it clear that experience-dependent, learned internal structures filter, select, and otherwise alter our perception and evaluation of sensory inputs. Such processes are so common that they seem only natural, and the excess of input beyond processing capacity makes them necessary. Two points, however, are of current relevance. First, since these internal structures select and value sensory input that is consistent with them, they create an exaggerated sense of agreement between the internal and external worlds. Second, since internal structures shape perceptual experience to be consistent with the structures themselves, they limit further alteration of brain structure by environmental input.

Consonance between Inner and Outer Worlds Is Pleasurable, Dissonance Is Unpleasurable

Observation of everyday life events and laboratory experiments make it clear that perceptual mechanisms operate in such a way as to select and value sensory input that is consonant with already existing internal structures, thus increasing the degree to which the external world is experienced as consistent with the internal one. Other studies have demonstrated a motivated component in actions that increase this consonance. People experience familiar stimuli as more pleasurable than unfamiliar ones, and experience conflict between their actions and their beliefs as unpleasant.

People experience things with which they are more familiar as more pleasing, merely on the basis of familiarity and independent of any qualities of the object. This was neatly demonstrated in the laboratory in a series of experiments by Robert Zajonc and colleagues. In one experiment, American college students were shown a series of Chinese characters with which they were initially completely unfamiliar [26]. They were told the experiment dealt with learning a foreign language and were instructed to pay close attention to the characters as each was displayed for a period of 2 seconds. Ten different characters were presented: two 1 time only, two 2 times, two 5 times, two 10 times, and two 25 times. Afterward the subjects were told the characters stood for adjectives and were asked to guess whether each character was positive or negative. The more often subjects had seen a particular character, the more often they assigned it a positive meaning. The frequency with which different characters were presented was balanced across subjects (e.g., the character presented 25 times to one group of subjects was presented only once to another group), unequivocally ruling out features of the stimuli themselves as the basis of the positive attributions.

Similar results are obtained when the stimuli are photographs of peoples' faces [26,27] or musical melodies [28,29] and the subjects are asked which person or melody they like more. The more they have seen the face or heard the melody, the more they like it, regardless of which particular face or melody was presented more often. When the stimuli are photographs of people's faces, the subjects also rate the people whose faces they have been shown more often as more like themselves, demonstrating the link between the internal structure (self) and familiar features of the external environment [30]. Moreover, the positive emotional response to previously seen faces or geometric patterns extends to averaged composite images of the faces and prototypes of the patterns even though the composites and prototypes had not themselves been seen before [31,32]. Since the composite is an internally created representational structure, this is even more direct evidence that consonance between internal structure and external stimulation is pleasurable.

Remarkably, the effects of mere exposure on a subjects' preferences are evident even when stimuli are presented so briefly that the subjects are not consciously aware of having seen them. For example, experimenters showed subjects pictures of ten irregular polygons, presenting each 5 times but for only 1 msec each time [33,34]. Each of the ten was then shown to the subjects paired with another irregular polygon that had not been previously presented. When the subjects were asked which one of each pair they had seen before, their accuracy was at the chance or guessing level. However, when asked which one of each pair they liked better, they picked the one they had seen before significantly more often than the one they had not. This indicates that the effects of familiarity on preference can be mediated by implicit memory systems.

Consistent with this, altering the size or orientation of an object [35] or the timbre of a melody [36] decreases recognition accuracy but does not decrease the effect of exposure on preference. Moreover, patients who have brain damage from excessive alcohol use [29] or Alzheimer's disease [28,37] that makes them unable to remember new information still show exposure effects on preference. While these studies indicate that implicit memory systems can mediate the effect of familiarity on preference, we cannot draw the conclusion that other brain memory systems do not also contribute to the pleasure associated with the familiar. The fact, though, that these effects can take place automatically and without conscious awareness makes them difficult to prevent, circumvent, or control.

Demonstrations of positive affective responses to the familiar are not confined to laboratory perceptual experiments. Perhaps most striking is the extremely strong association between preference for letters of the alphabet and the frequency with which each letter appears in common text, a correlation of 0.84 [38]. People like the letters best that they see most often. Similarly, we prefer the letters that make up our own names, although here the familiarity effect is supplemented by the investment we make in aspects of the world that we see as part of us [39].

In a test of the results of the laboratory perceptual experiments in a more natural setting, researchers printed the same set of Turkish words in college newspapers at two different American midwestern universities [40]. The words appeared daily for several weeks without explanation. Over time, some of the words were printed only once, others twice, and the rest 5, 10, or 25 times. Words presented frequently in one newspaper were presented infrequently in the other. Afterward, students on each campus were asked, in classroom settings and by mailed questionnaire, to indicate which words they liked better. As in the laboratory, the students preferred the words that had appeared in the paper most frequently, despite not knowing what the words meant or why they were in the paper.

The mere exposure or familiarity effect applies also to contact among people. In a clever experiment on the effects of repeated, incidental contact between individuals, researchers had one group of subjects, designated tasters, taste and rate each of eighteen liquids along a variety of dimensions (salty, bitter, etc.) as well as pleasantness. A second group of subjects, designated observers, also rated the pleasantness of each liquid based on highly standardized, 35-second, one-on-one, face-to-face observation of a subject tasting a solution. The subjects were rotated in such a way that each observer had contact with one taster once only, with a second taster twice, with a third taster 5 times, and with a fourth taster 10 times. The contact of tasters with different observers was similarly distributed. After this phase of the experiment, each subject indicated how personally attractive they found each of the four subjects with whom they had been paired. Individuals with whom the subjects had had more contact were rated as more attractive [41]. Or consider the example of a professor who, teaching an introductory course on "persuasion," had a mysterious student attend class every day enveloped in a big black bag with only his feet showing. The attitude of the other students toward the "black bag" was at first hostile, then curious, and finally friendly [27].

All theories of learning posit intensification of the neuronal representation (or capacity or probability for the neuronal representation) of objects as they are repeatedly encountered. Thus, as an individual becomes familiar with an object, its neural representation becomes more developed and more easily activated by subsequent exposures to that object or objects that share features of that object. The studies on the effects of exposure and familiarity on affective response indicate that sensory input that matches these more developed features of neuronal organization, or perhaps simply the activation of these developed features of neuronal organization, is experienced as pleasurable. A complementary line of research indicates that when sensory input disagrees with these established structures, the experience is not pleasing. These studies draw this conclusion from observing behaviors that minimize the discrepancy, rather than from direct statements by study subjects that they are unhappy or uncomfortable. Dissonance between established beliefs and expectations, on the one hand, and new information input on the other, is seen as creating a drive for its reduction [42].

The first line of defense against experiencing dissonance between internal structures and external reality is to avoid experiences that do not match one's internal structures. People seek out information that confirms what they already believe, and avoid information that does not, by selecting the people they associate with, the things they read, and the shows they listen to and watch. "This phenomenon of self-selection might well be called the most basic process thus far established by research on the effects of mass media. [It is] operative in regard to intellectual or aesthetic level of the material, its political tenor, or any of a dozen other aspects" [43, p. 138]. Despite such well-established and automatic processes, however, individuals still find themselves confronted by dissonant information in a variety of situations. Studies of individual responses in these situations demonstrate the discomfort associated with incongruities between internal structure and external reality.

The most common initial responses to such incongruities are to ignore, discredit, or forget the offending information. One study was done of students at Princeton and Dartmouth Universities after a particularly contentious football game between teams from the two schools, during which the nose of a Princeton player and then the leg of a Dartmouth player were broken [44]. After the game, the Princeton school newspaper accused the Dartmouth team of deliberately injuring their star player, while the Dartmouth paper reported that after an accidental injury to the Princeton player, the Princeton players began to deliberately play rough and dirty. A poll of students revealed that nearly 90% of Princeton students thought the Dartmouth team had started the foul play, and only 11% thought both had started it. In contrast, the majority of Dartmouth students thought both teams had started it. When students from each school were shown the film of the game and asked to note all the rule infractions they observed, the Princeton students noted twice as many infractions against Dartmouth as against Princeton, while the Dartmouth students noted an equal number of fouls against both teams. In other words, both groups of young adults at universities for outstanding students perceived the same events in different ways, each consistent with their preestablished views, and one (or the other, or both) ignored things that did not fit with those views.

Another study investigated the association between smoking habits and belief that a link had been proven between smoking and cancer [43, p. 154]. The survey was conducted in 1954, approximately 1 year after scientific research suggesting such a link had been reported and publicized. If only subjects who had not changed their smoking habits during that year are considered, any association between smoking habits and beliefs about the connection between smoking and cancer can be attributed to the effects of being a smoker. Over 4 times as many nonsmokers as heavy smokers thought the connection between smoking and cancer had been proven. Moreover, more light smokers than moderate smokers, and more moderate smokers than heavy smokers were convinced of the connection. Thus people for whom the new information conflicted with an existing belief to which they were behaviorally committed (and even addicted) tended to discredit the new information.

When making a choice among alternatives, individuals unavoidably create a situation of dissonance for themselves. At the beginning of the process information about the relative merits of the alternatives is considered. After one alternative is selected, information about the merits of the rejected choices conflicts with the selection. After making a decision, people work to reduce this conflict by forgetting and reinterpreting previously learned dissonant information, ignoring new conflicting information, and collecting additional information that is consistent with the decision already made. In a laboratory study of these processes, first-year students at Stanford University were told they were to be part of a two-person team to play against a single opponent [43]. They were then introduced to the two other players, one of whom would become their teammate and the other would be their opponent. One-third of subjects were given information to view about the two other players and then asked to choose one for their partner. Another third was given the information only after choosing their partner. The last third was given the information after indicating which of the two they would prefer to have as their partner, but were told the actual decision was not theirs because partners and opponents would be selected randomly. Since the experiment ended before any game was played, this last group of subjects was never told which individual was their partner.

The information presented to the subjects listed the strengths and weaknesses of both individuals to whom they had been introduced. The researchers measured the time each subject spent looking at all four types of information. Only the second group, those who had made their choice of partner before viewing the information, focused selectively on one category of information. These individuals spent most time studying the strengths of the person they had selected as their partner, thus choosing to spend the most time receiving stimulation that was consistent with an internally generated action. Later, the subjects were asked to recall the strengths and weaknesses of the two other individuals. By this time the first group of subjects had also selected a partner, and subjects from both groups one and two showed selective recall for information on the strengths of the person they chose to be their partner. Moreover, both groups actually reinterpreted as positive negative information about the person they chose. The third group of subjects, those who did not actually choose a partner, showed no biases in attending to, recalling, or interpreting the information. Thus, individuals use multiple means to minimize the discrepancy between an internally generated action that, once complete, becomes a feature of themselves, and perceived features of the environment.

Similar processes have been demonstrated in studies of decision making during everyday life. For example, researchers interviewed sixty-five adult men 4–6 weeks after they had purchased a new automobile, and sixty men from the same neighborhoods who had last purchased cars at least 3 years before [43, p. 51]. Prior to each interview, each subject was told the interview was part of a survey on readership of magazines and newspapers and was asked what magazines and newspapers he read regularly. The interviewer brought to the interview copies of the previous four weekly editions of the magazines and the previous seven daily editions of the newspapers the particular subject had said he read regularly. During the interview, the subjects were shown all the automobile advertisements that had appeared in the magazines and newspapers, and asked which they had noticed and which they had read at least in part. The control group of subjects, those who had not purchased a car within the past 3 years, provided a base rate for the percentage of car advertisements that men from this community read after noticing them. It turned out to be 37%. When men who had recently purchased a car came across advertisements for cars they had considered purchasing but did not, they behaved as did the control group and read 40% of them. This was close to the percentage of advertisements they read about cars they had not even considered purchasing (34%). However, these men read 65% of the advertisements they came across for the car they had recently purchased.

Before selecting a car to buy, reading advertisements for cars can be understood as an effort to gain information relevant to making a selection. The continued, selective reading of advertisements for a car that has already been purchased provides no such instrumental value. It does, however, help the individual experience a correspondence between the external world and a component of internal reality that has been created by the process of decision making and associated behaviors.

In the examples given here, decision making led to dissonance between information about the virtues of considered but rejected alternatives and the selected action, leading to a variety of behavioral and mental operations to strengthen the consonance between the action and the individual's perception of the world. Other studies show that if a decision contrary to existing internal beliefs can be coerced by social means, the resulting dissonance can lead to alteration of preexisting and now discordant beliefs. In one study, Princeton University students were asked to write essays that provided strong and forceful arguments in support of banning alcohol from the campus, even though they were against this position [45]. One group of students was then told that participation in the project was completely voluntary, and that they would be paid the same amount of money if they now decided that they did not want to write the essay. Further, they were asked to indicate whether or not their arguments could be given to a university policy committee by checking the appropriate box and signing their names. A second group of students was not given these choices, but instead was simply instructed to begin writing the essay. A third, control group was asked to write an essay providing arguments to allow the continued use of alcohol on campus, a position with which they agreed.

After writing the essay, the subjects were again asked to what degree they agreed or disagreed with the statement that alcohol use should be totally banned from campus. Only the first group of subjects, those who wrote a counter-attitudinal essay after being given the choice not to, showed a significant change in attitudes. Their attitudes about use of alcohol on campus became more consistent with the position they had espoused in the essay. The experimenters also measured electrical conductance of the skin, a peripheral measure of autonomic nervous system activity and physiological arousal. The group who wrote a counter-attitudinal essay after being given a choice not to, showed significantly more arousal than the other subjects. A similar study of male and female Adelphi University students asked to write counter-attitudinal essays in favor of instituting parking charges for students yielded the same results [46]. Students in the high-choice, counter-attitudinal essay condition showed greater attitudinal change and greater physiological arousal than those in the low-choice counter-attitudinal essay group or students in the pro-attitudinal essay group.

These same phenomena have also been demonstrated outside of the laboratory. For example, during World War II, the government of the United States made an effort to convince citizens to eat glandular meats that they were unaccustomed to consuming. Follow-up surveys indicated that after lectures only 3% of the attendees went home and served glandular meats. In contrast, 32% did so following smaller group discussions that concluded by asking individuals to raise their hands if they were now willing to try serving glandular meats [47].

In another study, researchers surveyed the attitudes of virtually all workers in a particular factory (over 2,000 people; ref. 48). Over the next 9 months, twenty-three of these workers were

made factory foremen, a position with management responsi-
bilities. After these workers had functioned as foremen for
between 6 and 12 months, their attitudes and those of forty-six
workers matched with the foremen on original attitudes and
demographic characteristics were resurveyed and compared. At
the time of this second survey, the workers who had become
foremen were much more likely than the comparison group to
have changed their minds in the direction of thinking that the
company was a better place to work than other companies, that
the union should have less to say in setting standards, that
things would be better without a union, and that the union-
supported seniority system was undesirable. These attitudinal
changes had the effect of decreasing the discrepancy between
the new foremen's attitudes and their behavior in their new
roles, and between their attitudes and those of their new management colleagues.

The workers accepted positions as foremen even though they
knew this would require them to carry out certain activities,
and represent certain positions, that were inconsistent with
some of their existing attitudes. According to dissonance theory,
they accepted the offer to become foremen because not to do
so would have created even greater conflict through incon-
sistencies with important other beliefs and affiliations; for
example, beliefs that as husbands and fathers they should maximize their income, or beliefs that as men in the general society
they should maximize their power and prestige. The workers-
turned-foremen then changed their attitudes about general
issues regarding management and unions in part so that their
attitudes would be consonant with the policies they espoused
and the actions they took in their new roles. They also changed
their attitudes so that they would be consistent with the views

of the other foremen and managers that constituted their new peer group.

Groups with which people affiliate in the workplace, or in religious or community organizations, do more than provide a passive venue in which like-minded individuals can receive pleasing, self-confirmatory input from one another. Such groups also actively increase the conformity of members' beliefs. Membership in a group becomes itself an internal feature of the individual member, and maintenance of views inconsistent with those of the group is dissonant. For example, if you are a member of a political party, it is uncomfortable to find yourself in agreement with the position of a rival party on a particular issue. More than this, individuals who espouse positions contrary to those of the group are chastised, devalued, and threatened with exclusion from the group. This phenomenon has been clearly demonstrated in the laboratory.

In one experiment, researchers brought subjects together for what they were told would be the first meeting of clubs being formed to discuss the problem of juvenile delinquency [49]. In the first meeting they were asked to discuss a case history written in such a way as to suggest that the particular juvenile delinquent needed love and kindness. Three research assistants posed as group members. One agreed with all the statements made by the others. The second began by expressing the view that harsh punishment was the best response to the juvenile's delinquent behavior, but changed his opinion to the group norm as the discussion progressed. The third assistant, however, began and persisted in the view that harsh punishment was best. By the end of the discussion, the subjects made derogatory comments about the recalcitrant stooge, and a number suggested he be excluded from the club.

The effects of group pressure in producing conformity in the beliefs of individuals in the group have also been dramatically demonstrated in the laboratory. In the best known of these experiments, groups of four to six subjects were shown a line and asked to indicate which of three additional lines was the same length as the first [50]. All but one of the subjects had been instructed to make the same erroneous selection. The single uninstructed subject was positioned in the group so as to hear the responses of most or all of the others before making his or her own choice. A large fraction of these subjects went along with the group selection even though the error was clearly evident. In a series of subsequent experiments, the subjects were asked to indicate which of two tones presented through headphones was longer [51, p. 162]. Each subject was alone in a listening booth, but conversation included on the stimulus tape led them to believe five other subjects were participating at the same time. These other "subjects" made the wrong selection on sixteen of the thirty test trials (i.e., they indicated that the shorter tone was the longer one), and this information was made available to the subjects before they indicated their own response. On over half of the test trials, the subjects were induced to identify the shorter tone as longer by having heard what they took to be assessments of other members of their cohort group.

In another of these studies, experimenters tape recorded critical comments from the other pseudosubjects and played these when the true subject made a factually correct choice on a stimulus trial despite having heard the erroneous choices of the pseudosubjects. After a subject made a choice that contradicted the responses of the pseudosubjects, the subject would hear laughter or the comment "Are you trying to show off?" This critical social feedback significantly increased conformity with erroneous responses on subsequent trials. It is of further interest that the frequency with which subjects could be induced by social pressure to make erroneous perceptual judgements was different in Norway and France. Apparently this tendency is itself shaped by environmental input during development.

What Happens When the External World Changes So Much That It Cannot Be Matched with the Internal World?

The world presents an immensely rich and varied stream of stimulation (information) to the individual. Internal neural structures are created that correspond to those aspects of environmental stimulation that are most commonly experienced by a particular individual. These structures then limit, shape, and focus perception on aspects of the information stream that are most like themselves. This increases the sense of correspondence between the external world and the internal one, and progressively limits the power of sensory stimulation to change the structures. Concordance between external stimulation and internal structure is experienced as pleasurable, and individuals preferentially place themselves in situations in which incoming stimulation is likely to be in agreement with their internal structure. When discordant information is encountered, that information is ignored, discredited, re-interpreted, or forgotten.

Stimulation in the form of other human beings is particularly important in shaping neural development, and there is a corresponding importance of concordance between a person's internal structure and interpersonal stimulation after development.

Individuals seek out social interactions and group affiliations that provide input that is consonant with their internal structures. They are loath to lose these group affiliations, and will alter their beliefs and perceptions in order to remain in agreement with the attitudes and behavior of other group members. Sometimes, however, individuals, and even groups of individuals, find themselves in situations that produce a discrepancy between internal and external structures that is too extreme to be amenable to any of these restorative actions. If a close fit between internal and external structures is necessary for human development and well-being, distress and dysfunction should be evident when changes in the external world exceed the potential of perceptual processes and self-alteration to readily reconcile the two. This is indeed the case.

As discussed in early chapters, infants have few internal structures and are highly adaptable to their environment. However, they have very limited ability to act on the environment so as to decrease discrepancies that arise between their internal structures and subsequent stimulation, and they clearly show stress in those situations. If there are abrupt changes in aspects of their mother that have begun to be incorporated into their internal representations, infants immediately show signs of distress and act in ways interpreted by observers as efforts to get the mother to resume her usual behavior. For example, if, at an experimenter's request, mothers do not participate in their usual interactive behavior with their infants, but instead are unresponsive and expressionless, their infants appear wary, increase the frequency of visual checking on the mother, and act as if they are attempting to bring the mother out of immobility [52]. In other words, they show signs of distress and attempt to make the social stimulation conform to the experience-derived expectation. If mothers are asked to feign depression by decreasing the affective quality of their facial displays, talking in a slow monotone, minimizing body movements, and refraining from physical contact with their babies, the babies increase their crying, back-arching and writhing movements, and serious, sober facial expressions and have shortened, truncated smiles or positive facial expressions [53]. If mothers simply turn their faces so as to present a profile rather than a head-on view of their faces to their infants, the babies decrease their smiling, vocalization, and physical activity [54].

When American, western European, and Japanese adults were asked to indicate the level of stress associated with different life events, death of a significant other, divorce and marriage were at the top of the list [55]. In each instance, the external world had changed so as to no longer match internal structures. The ranking of marriage just below divorce in the level of associated stress indicates that the disjunction between internal and external worlds that results from these major life events is one thing that makes them stressful, even when the event is desired and initiated by the individual. Adults are capable of making changes in their internal worlds to adjust to major changes in their external worlds, and many of these stressful adjustments eventually prove successful. These adaptations, however, are generally more difficult the older the individual, are experienced as unpleasant, can have negative physical and emotional consequences, are not always successful, and as a result are avoided when possible. The consequences and responses to one of the aforementioned stresses, death of a significant other, are so pronounced that they have been labeled a public health problem

[56] and studied extensively. Another major change in the external world results from immigration into a new cultural and physical environment.

Bereavement

The symptoms of grief and the process of mourning vividly reveal the effects of an abrupt disjunction between internal structure and external stimulation, and the time and effort necessary to recreate a comfortable consonance. The immediate symptoms of separation distress or bereavement include yearning for the lost individual; preoccuped thinking about the lost individual; symptoms of distress such as crying and sighing; depression with associated decreases in energy and activity; and dreams, tactile or visual illusions, or even hallucinations of the lost person [57,58].

Such symptoms seem so natural and expected under the circumstances that it requires an intellectual effort to distance oneself enough to appreciate what these symptoms indicate about the nature of human beings. In some instances the death of a spouse may raise real concerns about financial security for the survivor, even to the point of worry about stability of food and shelter, but examples abound where such concerns are absent and the injury to the bereaved is not instrumental in those senses but rather is what we call emotional or psychological. The emotional injury is generally and easily understood to be the loss of the deceased person him or herself. But just what has been lost? Concretely, and fundamentally, it is seeing, hearing, smelling, touching, and being touched by the other person. In other words, a large part of the interpersonal sensory environment that had become a large part of the internal representation of the external world is now gone. Moreover, in many instances the lost person also confirmed beliefs and ideas of the bereaved through regular conversation. The illusions a bereaved person may have of being touched by their lost partner, or the hallucinated experience of seeing him or her across the room, are like the simple sensory hallucinations experienced by individuals after extended sensory deprivation; both represent the neural activity of internal structures made excitable by the absence of customary and homologous sensory input.

In approximately 25% of recently bereaved individuals, symptoms of grief are so severe and persistent that it is considered pathological [59]. In these individuals, the nature and impact of the loss is even clearer. Thoughts about the deceased are frequent, intrusive, and distressing and are accompanied by intense feelings of yearning and longing. The feeling that part of oneself has died is common enough to be one of the diagnostic criteria for pathological grief, as is the feeling that one's worldview has been shattered [59]. Both social functioning and work performance are compromised; sleep is disturbed and self-esteem drops [59]. Individuals suffering from this extended grief syndrome search for the deceased spouse even while "knowing" full well that he or she is no longer alive, and at times attempt to recreate the deceased by assuming his or her behavior, characteristics, and even medical symptoms [59].

Hallucinated experiences involving the deceased, or assumption of their characteristics, are short-lived (if extended, they are pathological) attempts to address the sudden disjunction between the internal and external worlds by recreating the missing environmental stimulation. The successful restoration of harmony, however, requires the systematic and thorough restructuring of the inner world to match the now-altered outer world. In this "grief work," described by Freud in "Mourning and Melancholia" [60], the internal expectation of each important activity and situation in which the deceased was an important part must be restructured without the deceased. This undoing of the wide range of experience-induced neuronal interconnections that had linked representation of the deceased with those of multiple, still-present features of the environment, is a substantial undertaking. It normally takes about a year of effort to restore a reasonably comfortable equilibrium between internal structure and external reality.

The time necessary to mourn, and the negative impact on other aspects of functioning, indicate the magnitude of the problem created for an individual when a major part of their familiar environment is dramatically changed. The stressfulness of bereavement is further documented by the increased incidence of medical illnesses in the bereaved. Mortality is significantly higher among the bereaved for 2 years following the death of their spouse, compared with other adults of the same age and same demographic and health backgrounds [61], with studies indicating up to a tenfold increase in risk of death during the first year of bereavement [62]. There are increases in suicide, accidents, heart disease, cancer, tuberculosis, alcohol consumption, and cirrhosis [63].

Immigration

The experience of immigration provides another demonstration of the effects of a major disjunction between internal structure and external reality, although here an individual's immediate family and closest interpersonal relationships may remain intact while nearly everything else in their environment changes. There is a rich literature consisting of personal descriptions of leaving a homeland and accommodating to a new culture, and there are a few papers by psychoanalysts who have worked with immigrant patients.

Eva Hoffman was 13 years old when she, her parents, and her 9-year-old sister left Poland in 1959 to emigrate to Vancouver, Canada. Growing up as part of a Jewish family in Cracow after World War II, Eva was in many ways on the margin of Polish society and the object of more than a few exclusionary and critical comments by her peers. Yet this was her only world, external and internal, until she and her family set sail for North America. As she explained, "the country of my childhood lives within me with a primacy that is a form of love. . . . It has fed me language, perceptions, sounds, the human kind. It has given me the colors and the furrows of reality, my first loves. The absoluteness of those loves can never be recaptured. No geometry of the landscape, no haze in the air, will live in us as intensely as the landscapes that we saw as the first, and to which we gave ourselves wholly, without reservations" [64, pp. 74–75]. She recalls that walking around Cracow shortly before leaving "I burst into tears as I pass a nondescript patch of garden, which, it turns out, holds a bit of myself" [p. 88]. Standing at the railing of the ship on which she is to depart, "I feel that my life is ending" [p. 3]. "When the brass band on the shore strikes up the jaunty mazurka rhythms of the Polish anthem, I am pierced by a youthful sorrow so powerful that I suddenly stop crying and try to hold still against pain. I desperately want time to stop, to hold the ship still with the force of my will" [p. 4]. How clearly she describes the formative effects of the environment into which she happened to be born, the connection between her internal and external worlds, and the impossibility—in her situation—of keeping the internal world together with the external world by which it was shaped and to which it was matched.

On her third night in Vancouver she has "a nightmare in which I'm drowning in the ocean while my mother and father swim farther and farther away from me. I know, in this dream, what it is to be cast adrift in incomprehensible space; I know what it is to lose one's mooring. I wake up in the middle of a prolonged scream. The fear is stronger than anything I've ever known" [p. 104]. A short while later she wonders "what has happened to me in this new world? I don't know. I don't see what I've seen, don't comprehend what's in front of me. I'm not filled with language anymore, and I have only a memory of fullness to anguish me with the knowledge that, in this dark and empty state, I don't really exist" [p. 108]. Aware when she was leaving her motherland that part of herself was being left behind because she was losing the external match to her internal self, in the new world she feels loss, discomfort, terror, and confusion when she is surrounded by an environment that does not match the inner world she has brought with her.

While those mourning a deceased spouse may feel that part of oneself has died, this transplanted immigrant felt as empty as if she no longer existed. When encouraged by those around her to try and forget what she left behind, she wonders "Can I really extract what I've been from myself so easily?" [p. 15]. When she attempts to take in her new environment, the requisite internal structures are lacking or the old structures are obstructing. Her situation is like that of the kittens described in chapter 2 who were raised seeing only vertical and horizontal lines and as a result had few brain cells that responded to oblique lines. "The city's unfocused sprawl, its inchoate spread of one-family houses, doesn't fall into any grid of mental imagery, and therefore it is a strain to see what is before me. . . . Even on those days when the sun comes out in full blaze and the air has the special transparency of the North, Vancouver is a dim world to my eyes, and I walk around it in the static of visual confusion" [p. 135]. When she looks at others who are farther along than she in the adjustment to a new world, she worries that even for them "insofar as meaning is interhuman and comes from the thickness of human connections and how richly you are known, these successful immigrants have lost some of their meaning" [p. 143].

The change in language associated with many immigrations further disrupts the links between the self and others, and between internal neuropsychological processes and external social processes. Two days after their arrival Eva and her sister are taken to school and given new names: Eva for Ewa and Elaine for Alina. After the teacher introduced them to the class, mispronouncing their last name in a way they had never heard before, "we make our way to a bench at the back of the room; nothing much has happened, except a small, seismic mental shift. The twist in our names takes them a tiny distance from us. . . . Our Polish names didn't refer to us; they were us as surely us as our eyes or hands. Those new appellations, which we ourselves can't yet pronounce, are not us . . . make us strangers to ourselves" [p. 105]. Their original names were, of course, assigned to them by others. But the assignment of new names after years of hearing the original names in association with themselves, and the internalization of those names in important neural structures, is no small matter. Eva quickly learns English but "the words . . . don't stand for things in the same unquestioned way they did in my native tongue. For example, " 'River' in Polish was a vital sound, energized with the essence of riverhood, of my rivers, of my being immersed in rivers. . . . 'River' in English . . . has no accumulated associations for me. . . . It does not evoke. . . . The process, alas, works in reverse as well. When I see a river now, it is not shaped, assimilated by the word that accommodates it to the psyche. . . . This radical disjoining between word and thing is a desiccating alchemy, draining the world not only of significance but of its colors, striations, nuances—its very existence. It is a loss of a living connection" [pp. 106–107].

Experiences like Eva Hoffman's are not particular to immigrants from Eastern Europe, to immigrants leaving behind a home in which they felt comfortable and happy, or even to immigrants having to learn a new language. The writer Jamaica Kincaid had similar problems perceiving the world upon her arrival in New York city from Antigua when she was 17 years old. "I could not see anything clearly on the way in from the airport, even though there were lights everywhere" [65, p. 3]. And even though she had long dreamt of coming to New York to leave behind the persistent unhappiness of her life in Antigua, the newness of it all was deeply painful. "The undergarments that I wore were all new, bought for my journey, and as I sat in the car, twisting this way and that to get a good view of the sights before me, I was reminded of how uncomfortable the new can make you feel. . . . Everything I was experiencing . . . was such a good idea that I could imagine I would grow used to it and like it very much, but at first it was all so new that I had to smile with my mouth turned down at the corners. I slept soundly that night, but it wasn't because I was happy and comfortable—quite the opposite; it was because I didn't want to take in anything else" [p. 4].

Her experience the next day shows the difficulties immigrants can experience in the most mundane activities. "Seeing the sun, I got up and put on a dress, a gay dress made out of madras cloth—the same sort of dress that I would wear if I were at home. . . . It was all wrong. The sun was shining but the air was cold. . . . I did not know that the sun could shine and the air remain cold; no one had told me. What a feeling that was! How can I explain? Something I had always known—the way I knew my skin was the color brown of a nut rubbed repeatedly with a soft cloth, or the way I knew my own name—something I took completely for granted . . . was not so. . . . This realization now entered my life like a flow of water dividing formerly dry and solid ground, creating two banks, one of which was my past—so familiar and predictable that even my unhappiness then made me happy now just to think of it—the other my future, a gray blank. . . . I felt cold inside and out, the first time such a sensation had come over me" [pp. 5–6]. Even memory of the unhappiness she had emigrated to escape, because of its familiarity, was preferable to the pain of the disjunction between internal structure and external reality!

Psychoanalysts, describing their own experiences as well as those of their patients, argue for the generalizability of the experiences described by Hoffman and Kincaid [e.g., 66–69]. The objects and activities of everyday life are most profoundly missed: food; music; social customs; language; landscape; and street corners, houses, and cafes that made up their personal neighborhoods. "The anxiety consequent upon this culture shock challenges the stability of the newcomer's psychic organization" [70, p. 1052]. Attempts at restoration often include choosing to live in a climate and landscape that resemble the ones left behind, listening to one's native music, associating with fellow countrymen, and eating primarily food from their homeland. These can become pathologically, and maladaptively, fixed and extreme when some individuals eat only food of their native land, associate only with compatriots, speak only their mother tongue, and recreate their former domestic environments by furnishing their new residences with articles from "back home." As in bereavement, mourning is an essential feature of successful adaptation, with the systematic restructuring of the self based on the new instead of the old environment. The immigrant's ability to enjoy the movies, literature, and games—in other words, the popular culture—of the new country, is a sign of successful engagement in a process necessary for the development of a new self for the new environment. A "mixed" guest list for a dinner party at the immigrant's house, including natives of both the immigrant's new land and former land, is seen as a sign of successful adaptation [70]. In arranging such a party, the immigrant creates an external situation that reflects by homology the successful and harmonious creation of internal structures that integrate the new environment with sustainable features of the old.

Summary and Conclusion

During the first years of life, interpersonal and other sensory experiences create internal neural structures in their own forms. Once established, these internal structures alter a person's perception and experience to make them agree with the internal structures. People seek and create experiences that match their particular internal structures, and select information from the environment that most closely approximates these structures. When faced with information that does not agree with their internal structures, they deny, discredit, reinterpret, or forget that information. These maneuvers increase the sense of agreement between the internal and external worlds, but they also decrease the frequency of stimulation that conflicts with internal structures, thereby decreasing the likelihood of a change in these structures. From these behaviors we might deduce that people are motivated to maintain the consonance between their internal and external worlds. An interesting series of striking experiments leads to the further conclusion that what is familiar, that is, what matches past experience, is experienced as pleasurable merely for this reason.

Other people are particularly important features of the environment, and as a result, identification with social groups is of particular importance. The desire for consonance between one's self and the group will lead to extreme distortions of experience and behavior. Changes in the behavior of groups with which a person identifies, even without direct interpersonal contact, such as in the relationship between a sports team and its fans, can even lead to physiological changes within the individual.

This is not to say that internal structures in adults do not themselves change in response to changes in the physical and interpersonal environment. Indeed, when individuals occupy new social roles, their attitudes change to match those of their new peers. Such internal modifications, however, are less common and more difficult once internal structures are established. Moreover, when changes in the environment are great, the corresponding internal changes are accompanied by distress and dysfunction, and are only accomplished through prolonged effort. The experience of the bereaved and the immigrant provide examples of the pain, discomfort, and dysfunction occasioned by a radical disjunction between internal structure and external reality, and of the time and effort required to modify internal structures to match the altered environment. The next chapter examines the reactions and struggles associated with contact between distinct cultures with different belief systems and customs. At these points of contact, each culture creates for the other a radical disjunction between internal structure and external reality—other human beings who look, act, and believe in ways that are discordant with internal representations of human being itself, and with a host of internal structures derived from formative social interactions.

5 The Meeting of Cultures

In all the scorched and exotic places of the earth, Caucasians meet when
the day's work is done to preserve the fullness of their heritage by the
aspersion of alien things.

—O. Henry, *Cabbages and Kings*

Immigrants experience substantial problems moving from one
culture to another. Their homesickness and culture shock can
be so strong as to make the familiarity of even former unhappiness seem preferable to their current distress. Their process of
adjustment is fundamentally similar to recovery from bereavement. In both instances, the once largely harmonious fit between
internal structure and external reality has been disrupted by a
major change in external reality. In the case of bereavement,
there is loss of a single individual who was an important component of many aspects of the bereaved's world. In the case of
immigration, important interpersonal relationships may be preserved, but nearly all other aspects of the environment change.
The response, and the process of recovery, are similar—distress
and dysfunction followed by a prolonged, effortful restructuring
of the internal world to match the now-changed external
world.

Among the characteristics of our epoch in the development of the human species, none are more distinctive and significant than the meeting of different cultures and the establishment, perhaps for the first time, of a single human culture. The meeting, and possible integration, of cultures separate and unknown to one another for thousands of years results from improved means of travel, new modes of information storage and dissemination, and increased population. The radical disparities between internal structure and external reality occasioned by the discovery of different peoples and cultures have created distress for entire communities, and communal efforts to reduce the distress that at times have been violent. This distress and these efforts to reduce the stress are the topic of this final chapter.

The Origins of Differences in Culture and Language

One hundred and fifty to 200,000 years ago the first biologically human primate was born in northeast Africa. Among the characteristics that presumably distinguished this individual and her immediate evolutionary predecessors from earlier primate and protohuman species were an extended period of postnatal development and increased capacity for spoken language. One can only speculate about the rate at which language and culture developed during the childhood of the human species, and to wonder about the extent to which a single, original culture and language existed before groups of human beings separated and became isolated from the original group. Isolated populations of chimpanzees today have rudimentary and distinctive cultures, seen, for example, in the use of different tools and procedures for gathering ants from their nests [1,2], and prehuman hominid species made tools and are thought to have hunted in groups and communicated vocally. Thus, the initial group of human beings may then have inherited a primitive culture developed by its evolutionary predecessors. Until the advent of a developed language, however, this culture would most likely have remained rudimentary. As discussed in chapter 3, an individual develops competence in a language through imitation of the language he hears spoken. Moreover, the capacity for acquiring language diminishes dramatically after puberty. The original humans had to both develop a language and pass it on to their offspring. It is difficult to imagine it not taking many generations, and hundreds if not a thousand years to establish a language of any richness and complexity. It is likely that many small groups of individuals separated from the original group during this time, and the diversity of languages and cultures developed without there being an original of either. Other scenarios, of course, can also be imagined. Perhaps the protohumans who gave rise to the first humans had a more advanced language and culture than we imagine. Or, perhaps language and culture developed much more rapidly in one of the early groups of humans. This group may then have overtaken, conquered, or assimilated the others, thereby establishing the first universal language and culture, albeit on a much smaller scale than is happening today.

Whatever the truth in these speculations, there is no doubt that over the past 100,000 years human beings have spread around the globe, establishing many communities. These communities, like Darwin's finches, existed largely in isolation from one another for tens of thousands of years, and distinct languages and cultures evolved, leading to what Jane Goodall has called cultural speciation [3]. Some contact through war and trade existed between neighboring groups, but for most of human history this was limited to only the most proximate groups, and mountain ranges and water masses ensured total isolation of some groups from others for millennia. Indeed, until about 500 years ago, the peoples of the Americas knew nothing of the existence of, and were unknown to, the peoples of the rest of the world. Even within the small island land mass of Papua New Guinea, dozens of cultures developed in near isolation, separated by terrain that was difficult to traverse, and then by custom, mutual suspicion, and culture. It is true that trade, for example along the Mediterranean or the long silk and spice roads, did provide some contact among peoples of the extended African, Asian, and European land mass in recent millennia. Nevertheless although items from afar found their way into many cultures, the overwhelming majority of people still spent their entire lives within their immediate community. Even three generations ago most inhabitants of the technologically relatively advanced European land mass spent all but a few days of their lives in their own village. Travel as a pastime, even for the most educated and wealthy, is largely a phenomenon of the nineteenth and twentieth centuries.

The Extent of Differences among Cultures

The catalog of differences among the distinct human communities is extensive. Variance among languages is so great as to have eluded the efforts of scholars to arrange an orderly classification or identify overall patterns of development and interrelation. Even food consumption—what could be more basically biological?—has been shaped by different cultures in radically different ways. As one sociologist quipped: "Americans eat oysters but not snails. The French eat snails but not locusts. The Zulus eat locusts but not fish. The Jews eat fish but not pork. The Hindus eat pork but not beef. The Russians eat beef but not snakes. The Chinese eat snakes but not people. The Jali of New Guinea find people delicious" [4]. Similar variety is evident in customs of dress and the extent of the body exposed. Muslim women in some societies allow only their eyes to be seen; Orthodox Jewish women shave their heads when they are married and wear wigs rather than showing their own hair; American women expose their heads, faces, arms, and legs but cover their breasts and pelvic region; and Tasaday women of the Philippines conduct their daily activities unclothed [5]. Physical gesticulation and interpersonal physical space also vary; people from cultures nearer the equator stand closer to one another, hug and touch one another more, and gesticulate more when speaking than do individuals from more northern European and American cultures.

Beyond these more physical differences, differences even among relatively similar cultures have been documented in a variety of beliefs, preferences, and behaviors. One study, for example, identified the forty-five most successful plays produced in 1927 in Germany and in the United States, based on the number and content of critical reviews, the number of productions, and box office success. All ninety plays were summarized in English, with references to national origin eliminated (elimination of all such references was not possible in four German plays, so four American plays were Germanized). Nine judges, blind to country of origin, then assigned each play to categories for basic theme, age and gender of central character, ideological or personal level of action, and happy or tragic endings [6]. The plays popular in Germany were much more likely than the plays popular in the United States to have idealism (44 versus 4%), power (33 versus 2%), and the experience of being an outcast (18 versus 0%) as central themes. Plays popular in the United States were much more likely to have love (60 versus 31%) and morality (36 versus 9%) as central themes. The level of action in popular plays from the United States was rated as almost always personal rather than ideological (96 versus 4%), while the majority of the German plays were rated as ideological rather than personal (51 versus 47%). The central characters in the plays from America were female almost as often as male (15 versus 19%), and were more often youthful than middle aged or elderly (23 versus 13%). In contrast, the central characters in the plays from Germany were much less likely to be female than male (9 versus 33%), and were more often middle aged or elderly than youthful (19 versus 13%). Finally, the plays popular in Germany were 3 times as likely to have tragic or unhappy endings than the plays popular in the United States. These differences between plays popular in Germany and the United States reflect differences in the interests, concerns, fantasies, and aspirations of theatergoers in the two cultures.

Attitudes toward authority also vary among cultures, even if the individuals evaluated in each country are similar with regard to religion, gender, and socioeconomic status [7]. For example, 80% of American high school students thought it was right for a boy to run away from home if his father was cruel and brutal, but only 45% of German students and 46.5% of Filipino students concurred. Only 9% of American students thought older brothers had the right to give orders to younger brothers and obtain their obedience with force, while 23% of the German sample and 29% of the Philippine sample thought older brothers had these prerogatives. Eighty-four percent of the Americans and 67% of the Filipinos thought it was right for a soldier to refuse to obey the order of a superior to shoot an innocent military prisoner, but only 50% of the Germans thought such disobedience was justified.

Gendered roles in the work place differ dramatically, even among highly industrialized nations. For example, the percentage of managerial roles filled by women in 1993 was only 2% in South Korea and 17% in the United States, compared with 28% in Austria and 48% in Switzerland [8].

Marked differences in values, and associated actions, can persist between communities even when those communities are nearby and aware of one another. For example, consider the study of decisions made in the 1950s regarding the graveling of village streets and the construction of high school gymnasiums in two small communities (populations approximately 250) located within 40 miles of one another in nearly identical natural settings in the southwestern United States. One village had been settled 80 years before by Mormon missionaries selected and sent by church authorities. Missionary activities had long subsided and the village was a typical Mormon farming village. The family farm was the basic economic unit, but partnerships based on kinship and a village cooperative were also important parts of the economic structure. The other village was settled during the Depression of the 1930s by migrants from a neighboring state attracted by free land available for those willing to move to the area and establish farms. Each farm unit was operated by a nuclear family. A construction company that was in the area to build a state highway offered to gravel the streets of both towns. At a well-attended town meeting, the residents of the Mormon town each agreed to pay $20, which together with a larger sum from the owner of the local store, covered the expense of the gravelling. Apparently one man of importance locally initially objected, but was silenced when a much poorer man cited the Mormon values of progress and cooperation and pledged to pay $25 himself. The residents of the other village rejected a similar plan, and instead each of the operators of the town's business establishments paid to have the areas in front of their businesses gravelled, leaving the rest of the village streets a sea of mud in rainy weather [9].

A short while later both villages were offered money by the state government for materials and skilled labor to build a high school gymnasium if the town would provide the labor for construction. The Mormon village developed a plan in which each able-bodied man contributed 50 hours of labor or $50 to the project. Some resistance developed, and community leaders had to make sustained efforts to ensure compliance, but the project was completed successfully. At a town meeting, residents of the second village rejected a similar plan, and many expressed the view that they had to look after their own farm and family and could not come into town to work. Additional funds became available from the state to start the project, but even after the foundation was built and bricks delivered, community participation failed to materialize and the project remained a hole in the ground.

Finally consider cultural differences in premarital sexual behavior as reflected in out-of-wedlock birthrates and in the percent of marital first births occurring within the first 6 months of marriage [10]. Official local government statistics for the year 1955 indicate an out-of-wedlock birthrate of 0.9% in the predominantly Mormon Utah County in the state of Utah; 2.9% in Tippecanoe County, Indiana, another midwestern American community; and 6.6% in the country of Denmark. In order to determine the rate of births within 6 months of marriage, researchers compared marriage and birth records. Nine percent of births occurred within the first 6 months of marriage in Utah County, 9.7% in Tippecanoe County, and 32.9% in Denmark.

Importance of Differences among Cultures

Differences among cultures abound. It is a more difficult matter to determine in what ways and to what degree these differences are important. One thesis of this book is that the developing human brain shapes itself to its environment, and that the particular form of that environment (i.e., culture) is relatively unimportant. Indeed, the range of cultural variation despite a common human biological substrate is evidence that the particulars of cultural variation are not biologically important. A second thesis of this book, however, is that incongruities between the environment and the developed brain, incongruities introduced by significant changes in the environment, produce distress and dysfunction. The ability of a culture to minimize such disjunctions between internal and external worlds can thus be seen as a positive attribute, and some of the differences among cultures may be important in this regard. More important, according to this thesis, are the reciprocal disjunctions between internal and external worlds created by the meeting of different cultures. The experience of immigrants transplanted from one culture to another has been offered to indicate the difficulties inherent in such juxtapositions. Several additional lines of evidence also support this view.

Importance of Ceremonial Objects

The first of these additional lines of evidence is the importance people assign to the cultural, symbolic, or ceremonial aspects of objects above and beyond their instrumental value. Archaeologists distinguish ceremonial objects from those used in everyday tasks such as hunting or food preparation by the care and labor that went into their construction. Even peoples with technologically undeveloped cultures, for whom matters of subsistence occupied a large part of the day's activity, spent more time when making a ceremonial object than when making a utilitarian object. These ceremonial objects served only as the external manifestation of an internal belief, but for that very reason were particularly prized. The same is true on the community level, where the greatest collaborative efforts were devoted to the building of structures to represent belief systems.

European villages worked for one or two hundred years to create cathedrals huge in scale and adorned with sculpture, stained glass windows, and paintings. Or consider the controversy a few years ago surrounding the plan of the majority population of the town of Nazareth, Israel, to build a mosque adjacent to an important Christian pilgrimage site, the Basilica of the Annunciation [11]. "Why should anyone be upset with us?" asked a local Muslim councilman. "We are not building a dance hall or a casino. We are building a house of prayer. May our city be filled with blessing upon blessing." He well knew why his Christian neighbors were upset, as did the thousands of Muslims who "chanted prayers, threw fists into the air and set off firecrackers" as they laid the cornerstone for the new mosque. Christian churches were closed for 2 days in protest, turning away pilgrims who had traveled from abroad to visit the basilica. One Christian resident of Nazareth complained, "It's sad, though, shameful. We have the same language, the same culture [almost!], the same foods. We grew up together, neighbors. That's what hurts me, that these people I went to school with turned out to be fundamentalists. What if we tried to put a church in Mecca, their holy place?" (Of course, Christians have done one better in the past, building a cathedral within the great mosque in Cordoba, Spain, after taking that part of Spain by force.) Repercussions were even felt internationally, with the Vatican issuing a sharply worded criticism of the Israeli government's decision to authorize construction of the mosque.

Investment in Personal Environments and Objects

Similarly, the importance of personal environments is also readily apparent on the individual level during the course of everyday life. Consider the reaction of a teenager who within minutes of arriving on the campus of a college to which she might apply, announces to her dismayed parents "It's off the list. It doesn't feel like me." What would a convent, or some even more foreign institution feel like as a place to spend the next 4 years of her life? Or consider the lament of a 14-year-old boy trying to buy a pair of shoes. He was paralyzed with indecision. One brand was functional but the company had a bad record on environmental issues. Another pair appealed to him because they were "retro," but he was too young to wear such shoes and would feel like a "poser." Others were too high style, just not him. Of course, whatever pair he picked would soon feel like an appendage of himself and would have to be pried off after a year of daily wear before he could be persuaded to go through the painful process of selecting a new pair again.

Finally, consider the amusing but true story of the "ecotourist" from New Zealand on a recent trip to China. Ecotourists have an ideological commitment to maintenance of natural environments; particularly enjoy travel to places still in their pristine, natural state; and support tourism and travel with low environmental impact. While eating in the dining car of his train, the ecotourist noticed the Chinese cook throwing containers out of the train window as soon as they were empty. As this continued, the tourist became unable to contain himself and attempted, despite a complete language barrier, to convince the cook to stop the offensive behavior. As the failure of his effort at persuasion continued to be apparent, the tourist became increasingly agitated. After louder attempts at persuasion also failed, he tried to physically stop the cook from continuing to litter the countryside and eventually had himself to be restrained! Thus, we see that people demonstrate the importance of culturally specific aspects of their environments in the effort they devote to the construction of such objects (small and large), the offense they take at being confronted by religious displays of another culture, the efforts they make to find a microculture or clothes to match their view of themselves, and the agitation they feel by seeing behavior that conflicts with their ideology.

Troubling Fascination with Foreign Cultures
Further support for the importance of cultural differences can be found in the intense interest foreign peoples show in each other at the time of contact, an interest perhaps best characterized as a compelling preoccupation or obsession. As argued in chapter 3, sensory stimulation in the form of other human beings is particularly important in development of the human brain, and internal structures based on interpersonal experience

are an important part of cerebral organization. The amygdala, a brain structure particularly active in response to social-emotional stimulation, is more active in both black and white subjects when they view pictures of faces of people of the opposite color rather than of the same color as themselves [12]. Over the course of exposure to many such pictures, activation following same-color faces steadily decreased while activation associated with opposite-color faces remained high. For this and other reasons, evidence of seemingly totally different human beings could not be mentally filed in the same way as reports of a new species of butterfly, or even experienced with the same detached excitement or interest that might attend reports of the discovery of the skeletal remains of a large prehistoric creature.

Tales and texts from the French scientific expedition to South America in 1735 "circulated round and round Europe for decades, on oral circuits and written" [13, p. 18]. Kolb's account of southern Africa, published in German in 1719, was translated into Dutch in 1721, into English in 1731, and into French in 1741. Sparman's description of travels to this same region was published in Swedish in 1783, translated into English in 1784 (four editions) and into Dutch and French in 1787. Another account of the peoples of this same region, *Narrative of Four Voyages in the Land of the Hottentots and the Kaffirs* by Paterson, was published in English in 1789 and translated into French and German the following year [13]. The first edition of Mungo Park's 1799 English account of travels in the interior of Africa sold out in 1 month, and two more English editions, an American edition, and French and German translations were published by 1800. So monumental was the discovery of radically different and previously unknown peoples that the public appetite for such accounts grew steadily in Europe and the United States over the next 100 years, culminating in a relative explosion of productivity in the genre in the final decades of the nineteenth century [14, p. 108]. Newspapers financed the travels of explorers in exchange for rights to publish their reports, and hundreds of articles appeared in periodicals such as the *Atlantic Monthly, Harpers Weekly,* the *North American Review,* and *National Geographic* [14, p. 108].

While the very number of such publications indicates the Euro-American fascination with peoples of Africa, Asia, South America, and the various island states, the encounters reported in the publications indicate that the interest was reciprocal. Mungo Park reports arriving in an African village during a feast day and at a time when the residents were dancing. "When they were informed that a white man was coming into town, they left off dancing, and came to the place where I lodged, walking in regular order, two and two. . . . they continued to dance and sing until midnight during which time I was surrounded by so great a crowd as made it necessary for me to satisfy their curiosity by sitting still" [pp. 104–105, cited in 13, p. 80]. At another time his inspection by the Africans was more direct: "The surrounding attendants and especially the ladies were abundantly more inquisitive; they asked a thousand questions, inspected every part of my apparel, searched my pockets and obliged me to unbutton my waistcoat, and display the whiteness of my skin; they even counted my toes and fingers, as if they doubted whether in truth I was a human being" (p. 109, cited in 13, p. 82).

Euro-American preoccupation with other peoples was too great to be satisfied by the accounts of explorers and led to the importation of individuals from other cultures for display and inspection. The American Museum of Natural History, established in 1877, included permanent exhibits devoted to indigenous peoples of Asia, Africa, and North America among its displays of other animals, but felt no need to include similar exhibits on European or Caucasian Americans. The exhibits included skeletal remains (again analogous to the animal exhibits), and, at times, living human beings [e.g., 15]. Individuals were brought from Africa to Europe and exhibited at circuses and fairs. Most dramatic were the anthropological exhibits at the St. Louis World's Fair in 1904. Groups of indigenous peoples from North America, South America, Japan, the Philippines, and Africa were brought to the fairgrounds and housed for exhibit in re-creations of their villages, residences, and habitats. The Filipinos alone comprised a group of 1,200 individuals living in a 47-acre exhibit [15,16]. The fair in general was a huge success, and the anthropology exhibits were one of its main draws. The fair had more than 18 million visitors [15, p. 123], and although it is unclear how many were repeat visitors, it is an astounding number given that the population of the United States was only 76 million in 1900 [17]. When Ota Benga, a young Pygmy man, was brought to New York City after the fair and housed temporarily in the Monkey House at the New York Zoological Park, 40,000 people came to see him in a single day. The *New York Times* reported that "nearly every man, woman, and child of this crowd made for the monkey house to see the star attraction in the park—the wild man from Africa. They chased him about the grounds all day, howling, jeering and yelling. Some of them poked him in the ribs, others tripped him up, all laughed at him" [quoted in 15, p. 185].

Perhaps there were a few among the crowds in St. Louis and New York whose interest in the strange peoples was the potential for their own economic advancement: far-sighted souls who came to gain information relevant to appropriating the wealth of these peoples' homelands or benefiting from the cheap labor their compatriots might supply. For the vast majority of the thousands and millions of visitors, however, the attraction was to see, and to begin to deal with, the radical and undeniable fact that people existed who looked and acted so differently from the people they had seen and lived with all their lives. And what better way to begin dealing with this information than when the individuals to be considered were kept separate by exhibition boundaries, presented in a context of fun and fantasy, and clothed in the procedures of familiar scientific inquiry and observation. As two scholars of this period of travel writings and cross-cultural encounters have noted, European culture had an "obsessive need to present and represent its peripheries and its others continually to itself" [13, p. 6], and "of all the world's distant places, Africa may ultimately have been of least concern to most Americans when it came to real and concrete questions of economy and invested interest. But the continent held a place of disproportionate interest in the geographic imagination" [14, pp. 116–117].

While irresistibly drawn to inspect these other humans, exposure to the intercultural differences was unsettling in ways similar to other disjunctions between internal structure and external reality. As indicated earlier, Mary Pratt coined the term *contact zone* to refer to "the space in which peoples geographically and historically separated come into contact with each other and establish ongoing relations, usually involving conditions of coercion, radical inequality, and intractable conflict" [13, p. 6]. Eighteenth- and nineteenth-century travel writers often described the foods and customs of foreign cultures as strange and revolting [13, p. 44], and the people themselves as dirty, and even being unable to cleanse themselves [14, p. 111]. J. M. Coetzee explicitly "attributes the widespread vilification of the Hottentots in the writings of seventeenth and eighteenth century Europeans to frustration at the Khoikhoi failure to fulfill anthropological and economic expectations" [quoted in 13, pp. 44–45].

These tensions are captured clearly in the ever-more negative connotations of the word "barbarian" from its initial use in the Greek of Herodotus to its meaning in modern English [18]. Herodotus's inquiry into the origins of hostility between the Greeks and Persians led him to visit and describe the belief systems, arts, and everyday practices of neighboring non-Greek societies. He called these societies barbarian, a word that in the Greek of his day meant people whose language, religion, ways of life, and customs differed from those of the Greeks themselves. Later Greeks used the term to mean "outlandish, rude, or brutal." Once incorporated into Latin, it came to mean "uncivilized" or "uncultured" rather than "of different culture." The *Oxford English Dictionary* describes its original meaning as "one whose language and customs differ from the speaker's" but then provides the contemporary definition as "a rude, wild, uncivilized person."

The conflict is all the greater when foreign customs present themselves in one's own homeland. No less an international man than Henry James, a student of the subtleties of cultural difference and a prodigious intellect, described his own deep discomfort upon returning to New York City in 1904 after 20 years in Europe. He visited "the terrible little Ellis Island" [19, p. 84] and saw the "visible act of ingurgitation on the part of our body politic and social, and constituting really an appeal to amazement beyond that of any sword-swallowing or fire-swallowing of the circus" [p. 84]. He asserts that

The action of Ellis Island on the spirit of any sensitive citizen who may have happened to "look in" is that he comes back from his visit not at all the same person that he went. . . . He had thought he knew before, thought he had the sense of the degree in which it is his American fate to share the sanctity of his American consciousness, the intimacy of his American patriotism, with the inconceivable alien; but the truth had never come home to him with any such force. . . . it shakes him . . . to the depths of his being . . . [creates a] new chill in his heart . . . [like a] person who has had an apparition, seen a ghost in his supposedly safe old house. Let not the unwary, therefore, visit Ellis Island. [pp. 84–85]

In New York he felt the "affirmed claim of the alien, however immeasurably alien, to share in one's supreme relation was everywhere the fixed element, the reminder not to be dodged" [p. 85]. He felt further that

The idea of the country itself underwent something of that profane overhauling through which it appears to suffer the indignity of change. Is not our instinct in this matter, in general, essentially the safe one—that of keeping the idea simple and strong and continuous, so that it shall be perfectly sound? To touch it overmuch, to pull it about, is to put it in peril of weakening; yet on this free assault upon it, this read-justment of it in *their* monstrous, presumptuous interest, the aliens, in New York, seemed perpetually to insist . . . so that *un*settled possession is what we, on our side, seem reduced to—the implication of which, in its turn, is that, to recover confidence and regain lost ground, we, not they, must make the surrender and accept the orientation. We must go, in other words, *more* than half-way to meet them; which is all the difference, for us, between possession and dispossession. This sense of dispossession, to be brief about it, haunted me so. [p. 86]

He envied "the luxury of some such close and sweet and *whole* national consciousness as that of the Switzer and the Scot" [p. 86]. When describing his reactions to seeing the large number of immigrants in New York itself, he explains that "as you go and come, the wonderment to which everything ministers and that is quickened well-nigh to madness, in some places and on some occasions, by every face and every accent that meet your eyes and ears . . . fed thus by a thousand sources, is so intense as to be, as I say, irritating. . . . the facts themselves loom, before the understanding, in too large a mass for a mere mouthful: it is as if the syllables were too numerous to make a legible word" [pp. 120–121].

Nonviolent Efforts to Deal with the Distress Occasioned by the Meeting of Cultures

Denial and Distortion to Fit Existing Beliefs

Given the discomfort occasioned by the colossal discovery of foreign peoples and cultures, and the resulting, radical disjunction between important aspects of internal structure and external reality, it is not surprising that we see large-scale employment of the same response mechanisms described previously in individuals confronted with sensory input that fails to match their internal neurocognitive structures. According to scholars of the academic and popular travel and exploration literatures of the eighteenth and nineteenth centuries, two mechanisms in particular were widely and systematically used: denial of the offending information or distortion and manipulation of the information and peoples to fit into preexisting scientific or religious belief systems.

In part, the denial of aspects of the foreign cultures was due to the absence of the internal structures necessary for perceiving them. Just as kittens raised in a visual environment with stripes that move only left to right have fewer brain cells that respond to movement from right to left, just as Eva Hoffman was unable to see the features of her new Canadian landscape, or Jamaica Kincaid was unable to make out the things she saw on the way into New York City from the airport, Egyptian markets, African villages, and immigrant communities in New York City are all described as full of confusing, bewildering swarms of unindividuated crowds [e.g. 14, p. 125]. And in part, the denial was a collaborative deployment of avoidant, repressive, and forgetting mechanisms described by psychoanalysts in their offices and experimental psychologists in their laboratories. Even particular features of new cultures reported by the earliest explorers disappear from later accounts [13, p. 52]. "The landscape is written as uninhabited, unpossessed, unhistoricized, unoccupied. . . . an asocial narrative in which the human presence, European or African, is absolutely marginal" [quoted in 13, p. 51]. Indigenous peoples are described as "doing nothing" in early writings, and until they were studied by anthropologists who lived among them, were considered to be without culture, without professions, without laws, and without institutions [13, pp. 44–45; 14, pp. 110, 111]. When the efforts of European missionaries to convert foreign peoples to Christianity failed, the foreigners were described as being incapable of religion [13, pp. 44]. Descriptions of new places listed the features of European towns and landscapes that were not present rather than the foreign features that were present [14, p. 111]. Photographers of the immigrants who filled the streets of New York with the smells, sounds, and dirt that so unsettled Henry James went to great effort to photograph them one or two at a time, cleansed of the details of difference [16, p. 222].

More common and more problematic than this denial were efforts to select and distort the perception of foreign peoples and their cultures and manipulate both to bring them into agreement with the existing belief systems and structures of the perceiver. On the European side of these intercultural encounters, first religious, and then religious and scientific belief systems were applied to the task of incorporating the newfound peoples and cultures. For example, when emigrants from England settled in the region of North America that they named "New England," they believed the Native Americans they encountered were the lost tribes of Israel described in European religious texts. The Englishman John Hanning Speke devoted much of his widely read 1863 report "Journal of the Discovery of the Nile" to describing the primitive people he encountered, concluding that they were the descendants of Noah's son Ham and "a strikingly existing proof of the Holy Scriptures" [quoted in 20, p. 51].

By the later half of the nineteenth century, Europeans and Americans brought scientific as well as religious belief systems to the task of dealing with the existence of foreign peoples. Scientists were dispatched to measure the newly discovered people in every which way, and individuals were brought to Europe for similar inspection in scientific laboratories. When the new technology of photography became available, objectified images of other peoples were created for analysis and illustration [16]. The conclusions of these scientific inquiries were consistent and foregone; based on Darwin's theory of evolution, which dominated scientific thinking about living things by the mid-nineteenth century, the newly discovered people were considered evolutionary precursors of Euro-Americans [14, chap. 4]. In some accounts, the scientific observations were seen as establishing that the newly discovered people were not human beings at all, but a distinct species filling the evolutionary space between men and monkeys. In others, the different peoples of the world were organized into a developmental or evolutionary hierarchy of groups, invariably with Europeans as the most advanced stage of development. The anthropological exhibits at the St. Louis World's Fair again provide a clear example. W. J. McGee, head of the anthropology department at the fair, explained in an essay entitled "The Trend of Human Progress" that human beings "are conveniently grouped in the four culture grades of savagery, barbarism, civilization, and enlightenment. . . . the two higher culture-grades [are]—especially the Caucasian race, and . . . the budded enlightenment of Britain and full-blown enlightenment of America" [quoted in 16, p. 278]. The exhibitions of foreign peoples at the fair were arranged to illustrate this progression. The same views were found everywhere, in school textbooks and scholarly and popular texts of the period [14, chap. 4].

The Sad Case of Captain James Cook

The human response to the discovery of foreign human beings seen in the Europeans was similarly evident in the response of indigenous peoples to the Europeans. "The Melanesian literature on 'first contacts,' often richer than the Polynesian because the events were more recent, confirms this disposition to interpret the intrusive coming of Europeans in ways consistent with the people's own cosmological schemes—including even the failure to remember the first white man as such, since they were never men" [21, p. 179]. Perhaps the most dramatic of the many examples of this are the events of James Cook's arrival, and eventual murder, in Hawaii [21]. Cook chanced to arrive off the coast of Hawaii at the start of the annual 2-week-long celebration of the return of the deity Lono. Cook's ship first approached land and anchored in the bay at the town of Kealakekua, which legend had was the home of Lono when he had been a man and the place to which he was expected to return. People on shore had apparently followed Cook's ship as it progressed toward Kealakekua, because 10,000 were there to greet him, several times the normal population.

When the ship entered the bay, the reception was extraordinary. Journals kept by those on board noted that they were soon "amidst an innumerable number of canoes, the people in which were singing and rejoicing all the way" and "all the shore of the bay was covered with people" [quoted in 21, p. 47]. Cook noted that no weapons were seen, and that the canoes were full with pigs, fruits, and sugar cane. Great numbers of natives climbed on board Cook's ships, and on the ships, in the water and on the shore, people were dancing, singing, clapping, and jumping up and down. When Cook came ashore in Kealakekua, and subsequent places along the coast, he was escorted by priests who called him Lono, wrapped him in special cloth, brought him to temples, and fed him special foods not offered to others. He was always induced to sit with both arms extended straight out from the sides of his body, supported horizontal to the ground by a priest on one side and by Cook's lieutenant on the other, thus assuming the position in which Lono is depicted in pictures and carvings.

In a scholarly tour de force, Marshall Sahlins aligned the Hawaiian and English calendars of 1779 and compared the dated entries in the journals of the English sailors with the documented sequence of feasts, rituals, and observances of the annual 2-week Hawaiian celebration. The changing nature of food offerings brought to the ships, and the sequence of religious and military officials visiting the ships, correspond precisely with the highly scripted progression of the holiday rituals, including even wrapping Cook in a ceremonial cloth and sacrificing a small pig in front of him. Cook's lieutenant observed that "this ceremony was frequently repeated [for Cook] during our stay at Owhyhee, and appeared to us, from many circumstances, to be a sort of religious adoration. Their idols we found always arrayed with red cloth, in the same manner as was done to Captain Cook, and a small pig was their usual offering to the Eatooas [akua, 'gods']. Their speeches, or prayers, were uttered too with a readiness and volubility that indicated them to be according to some formulary" [quoted in 21, pp. 49–51].

As the holiday ended, Cook's ships sailed on, apparently without knowing the cause of their warm and elaborate welcome. Misfortune struck, however, when a storm damaged the mast on Cook's ship, causing him to return to Hawaii for repairs. The Hawaiians were apparently confused and frightened by the return of Lono so shortly after his departure, and the response to Cook and his men was in marked contrast to their previous welcome. English journals indicate that the Hawaiians were not nearly as friendly as before, seemed fearful of Cook's intentions, and unable to understand or accept that a problem with the Englishman's ship had caused them to return. When some Hawaiians took one of Cook's launches, Cook decided to take the Hawaiian king hostage as a means to secure its return. The king, perhaps resigned to being carried off by Lono, is reported to have followed Cook passively, until his wife implored him not to go. The king's supporters then turned on Cook, killing him and carrying off his body, which it seems was then burnt in sacrifice and divided among the Hawaiian leaders according to their status.

More recent accounts of first contacts between Euro-Americans and isolated island communities all reveal similar efforts to perceive the Euro-Americans, not as different human beings, but rather as components of the existing indigenous belief systems.

A New Guinea culture that included a sky world in its cosmology thought Europeans were powerful spirits from the sky world, red in skin color because of their proximity to the sun. Others thought the Europeans were ancestral spirits from the land of the dead. When the Europeans washed river gravel looking for gold, the local people understood them to be looking for their own bones, which had been thrown into the river. Despite the fact that some brave warriors examined the Europeans at night and reported to the contrary, the belief that the Europeans turned into skeletons at night was widely held [21].

Violent Responses to the Meeting of Cultures

When contact between foreign peoples is sustained, simple distorted perception alone is often insufficient to manage the distress associated with the disjunction between internal structure and external reality. New information about the strangers emerges that is inconsistent with the distorted accommodation to existing belief systems, and the strangers act in ways that are inconsistent with the roles and characteristics assigned them (e.g., Cook's quick return to Hawaii). Confronted by these added insults to equanimity, each group attempts to induce or force the other to act according to the roles and characteristics assigned them, in a manner analogous to the role of projective identification in interpersonal interactions described in chapter 3. On the intergroup level, these efforts can contribute to large-scale violence and calamity.

Rwanda

Belgian influence on the peoples of Rwanda is a particularly tragic example [20]. Precolonial Rwandan society was an oral culture, and no reliable history of the region during that time exists. It is known, though, that the region was first settled by a cave-dwelling pygmy people, the Twa, who were then vastly outnumbered by immigration of the Hutu and Tutsi people. While quite different physically, "with time, Hutus and Tutsis spoke the same language, followed the same religion, intermarried, and lived intermingled, without territorial distinctions, on the same hills, sharing the same social and political culture in small chiefdoms . . . some of [the chiefs] were Hutus, some Tutsis; Hutus and Tutsis fought together in the . . . armies; through marriage and clientage, Hutus would become hereditary Tutsis, and Tutsis could become hereditary Hutus. Because of all this mixing, ethnographers and historians have lately come to agree that Hutus and Tutsis cannot properly be called distinct ethnic groups" [20, p. 45].

To be sure, the society had its inequalities and social roles, some of which were related to continuing distinctions between Hutus and Tutsis, and the distinction remained a feature of the culture. However, when Europeans explored Africa in the second half of the eighteenth century, the government of Rwanda was centralized, with a well-developed military, political, and civil structure, and the leader, revered as an absolute and infallible divinity, was the latest heir to a dynasty that had led the country for several hundred years. Rwanda had a reputation for "ferocious exclusiveness"; explorers were prevented from entering the country by force and the first white man is thought to have entered Rwanda only in 1894 [20, p. 54]. The missionary, Monsignor Louis de Lacger, wrote in his 1950s history of Rwanda that "One of the most surprising phenomena of Rwanda's human geography is surely the contrast between the plurality of races and the sentiment of national unity. The natives of this country genuinely have the feeling of forming but one people. . . . the ferocity of this patriotism is exalted to the point of chauvinism" [quoted in 20, pp. 54–55]. Another missionary wrote that the Rwandans "were persuaded before the European penetration that their country was the center of the world, that this was the largest, most powerful, and most civilized kingdom on earth" [quoted in 20, p. 55]. Lacger added that "There are few people in Europe among whom one finds these three factors of national cohesion: one language, one faith, one law" [quoted in 20, p. 55].

According to the analysis of Philip Gourevitch [20], the massacre in 1994 of 800,000 Tutsi Rwandans by Hutu Rwandans was the result, at least in part, of the unsettling and division of Rwandan society by imposition of a European myth on the Rwandan people. Belgian colonial potentates, together with their Roman Catholic missionary compatriots, understood Rwandan people according to the Hamitic myth propagated by Speke in 1863, and the associated myth of Tutsi superiority. Speke saw in most Africans the progeny of Noah's son Ham, cursed by Noah because Ham had stared upon Noah's nakedness. "Contemplating these sons of Noah," Speke wrote, "as they were then so they appear to be now" [20, p. 51]. Speke noted further that some Africans were taller and sharper featured than the others. These people, who more closely resembled the Europeans with whom he was of course very familiar, Speke concluded were a superior race descended from the biblical King David. Accordingly, the colonial and church leaders 70 years later "set about radically re-engineering Rwandan society along so-called ethnic lines" [20, p. 56]. Hutu leaders were replaced by Tutsis in even the local administrations that had provided Hutus with some autonomy and empowerment. The Belgians and the church replaced the Rwandan leader, who they found too independent, with another who converted to

Catholicism and renounced his divine status. Ethnic identity cards were issued, labeling each Rwandan as Hutu, Tutsi, or Twa. Sixty years of Belgian and church-supported Tutsi domination of the Hutus ensued. Large numbers of Rwandans converted to Catholicism; school books presented the doctrine of Tutsi superiority; and the idea of a collective national identity was replaced with the European myth and Rwandan reality of Tutsi superiority and dominance. In 1994 this de- and reculturated nation exploded with Hutu anger and its rivers filled with the bodies of Tutsis.

The European belief system did not require that the Hutus attack the Tutsis, and other factors were probably important in starting the violence. Jared Diamond, for example, suggests that pressures of a population density ten times that of neighboring Tanzania, and insufficient farm land to support the population, were important [22]. But the Europeans did coerce the Rwandans to abandon many of the beliefs that had formed the basis of Rwandan culture and society and to act in ways that conformed to the fantastical belief system that the Europeans had used to reconcile the existence of the strange people of Africa with their own notions of human beings. In so doing, deep animosity was created between Hutus and Tutsis that contributed to the violence. Gourevitch, reflecting on his study of Rwanda, succinctly notes "Like all of history, it is a record of successive struggles for power, and to a very large extent power consists in the ability to make others inhabit your story of their reality" [20, p. 48]. Imposition of the European story upon the Rwandans had sad consequences for the Rwandans. It is the neurobiological imperative to maintain consistency between internal structure and external reality that fuels the struggle to control the story, and this struggle often contributes to making the contact zone where previously separate cultures meet a place of coercion, radical inequality, and intractable conflict.

Coercion of one people to assume appropriate roles in the preexisting belief system of another people is, however, only a temporary and limited solution to the distress created by the meeting of foreign cultures. It serves to shore up the initial effort to perceive the different human beings in a way that is consistent with fundamental beliefs of the perceiver, rather than changing those established beliefs. In many cases, however, the foreign culture is too broadly different to be accommodated by such efforts, and/or members of one culture continue to act in ways inconsistent with the fundamental beliefs of the other. Here, then, the effort becomes elimination of the offending perception. Missionary and military forces are unloosed and the contact zone becomes the place of a death struggle between cultures. Indeed it is the dual use of missionary and military force that characterizes the European invasion of Africa and the Americas, a perplexing combination that is at least in part explained by a common utility in eliminating the offensive presence of difference.

A Methodological Note

Multiple factors can contribute to lethal conflict between different peoples and cultures. Competition for natural resources and export markets can be important. By the nineteenth century, the modern European state itself, and its needs for relative homogeneity among its constituents, contributed to the effort to eradicate indigenous cultures that the expanding state encountered [13]. Before that, intercultural incompatibilities in instrumental interactions with the environment, such as use of land for farming versus hunting, or establishment of permanent

residences on private property versus nomadic socialist communities, contributed to the elimination of Native American
people and cultures in the Northeast United States [23]. The
neurobiological antagonism to difference, and the associated
pressure to eliminate strange and foreign people with different
ideologies, is yet another fundamental factor that can contribute to violent conflict. Sometimes ideological differences are so
closely linked with other factors that it is impossible to separate
the two. For example, American and English colonist beliefs
about the proper relationship between people and the environment were incompatible for both resource management and
neuropsychological well being. In some instances, ideological
differences are not the primary motive for conflict but they
obstruct efforts to resolve the conflict by peaceful means. In
other instances, ideological differences may be a primary source
of conflict.

Data to support this assertion are not as clear-cut as the data
presented in chapters 2 and 3 that support the arguments for
environmental shaping of brain development, or the data presented in chapter 4 that support the assertion that adults work
to make their environments match their established inner neuropsychological structures. The problem of providing empirical
support for assertions about the causes of complex social phe-
nomenal is not limited to the present argument for the importance of ideological conflict in social violence, but applies
equally to assertions about other possible causative factors. In
general, support for the role of other causative factors is based
on theories derived from analyses of past social events. The
value of each theory is developed through its apparent ability
to help us understand other social and historical events, and
once its value is established the theory is applied to current
events. The argument offered here follows similar steps but from

a different starting point. Rather than being derived from analyses of historical events, the theory advanced in this book is based on scientific laboratory experiments. Its value has been demonstrated by additional laboratory experiments and by application to psychological and social problems seen in individuals or small groups. We now consider its utility in helping us to understand past and current major conflicts between cultures. In so doing, we will pay close attention both to what people did and why they said they were doing it. For this purpose, some rather lengthy edicts and statements are reproduced. Even in those that have been translated, the content and message is clear.

The Albigensians and the Inquisition

Consider the Inquisitions and the Crusades, violent campaigns that left millions dead in a 400-year effort by the Roman Catholic Church of the time to eliminate competing belief systems within Europe itself as well as on its border with the Muslim world. These campaigns are notable for their powerful, persistent, and explicit ideological foundations, but also for the fact that they reflect a church with a very different character and understanding of Christianity than that of the Roman Catholic Church today. Indeed, the role the Catholic Church plays today in promoting understanding among different faith groups demonstrates how cultural institutions can change over time.

The Inquisition, a system of tribunals established to judge and punish those accused of heresy, was the outgrowth of a crusade proclaimed in March 1208 by Pope Innocent IV against Raymond IV, the Christian count of Toulouse, who governed southern France. Raymond was a cousin of the king of France and brother-in-law of the kings of England and Aragon. What distinguished him and his land were not ethnic or general cultural differences, but the Albigensians, members of the Cathar Church, whose home was in his dominion. Although a Christian church, Cathar doctrine differed from that of the Roman Catholic Church in fundamental aspects of belief, differences that in addition to their religious significance led to differences in public behavior and life-style. Despite the many connections with the broader European community, the differences between the Albigensians and this broader community were substantial enough for the Polish historian and essayist Zbigniew Herbert to conclude that "in the South of France there existed a separate civilization, and that the Albigensian crusade was a clash of two cultures. The defeat of the duchy of Toulouse is one of the catastrophes of humanity, like the destruction of Cretan or Mayan civilizations" [24, p. 107]. After missionary efforts to convert the Albigensians to Roman Catholic doctrine failed, the Catholic Church granted "an indulgence of two years to those who shall make war on "the Cathars," citing their destruction of Catholic churches as well as their heretical beliefs [24]. In June 1209 an army of 200,000 French, German, and Dutch men marched on the duchy, led by bishops, archbishops, dukes, counts, barons, and knights [24, p. 109]. Over 15 years of fighting followed, with estimates of the dead as high as one million [24, p. 115].

The treaty concluding the surrender of most of the Albigensian homeland included a provision to pay two marcs to anyone apprehending a heretic, for although the military contest was essentially decided, the full elimination of heretical beliefs was a matter of continued effort. The Inquisition was born in the forty-five articles for identifying, judging, and punishing here-tics produced by the synod at Toulouse in 1229. Among the articles [as quoted in 24, pp. 116–117] are:

In each parish bishops will appoint one priest and three—or more if necessary—laymen of impeccable reputation, who will swear to search for heretics in their parish with perseverance and faith. They will meticulously search all suspicious houses, rooms and cellars, and even the most secret corners. Upon finding heretics, or people giving them support, shelter or aid, they should undertake appropriate measures to prevent their escape, and also notify the bishop, the lord or his representative. A lord's deputy, who does not search the places suspected of being heretical meeting-places with sufficient ardour, will lose his position without redress. Everybody has the right to search for heretics on his neighbour's land . . . Also the King can chase them on the land of the Count of Toulouse, and vice versa. All adult Catholics will take an oath in front of their bishop to keep their creed, and to track down heretics with all the means at their disposal. This oath is to be renewed every two years.

Provisions were made for the heretics to renounce their heresy and join the Catholic community, but those who refused—and there were many—were burned alive. The choice is made clear in the refrain from a troubadour's poem:

If you do not believe, turn your eyes to the flames which
burn your companions.
Answer in one or two words only—
You will either burn in this fire, or join us.

Even the buried bodies of dead heretics were exhumed from the cemeteries to be purified by fire. Those heretics who would renounce their heresy and embrace Catholic doctrine could be part of the community. However, even those who were already removed from the community by death were exhumed and burned to protect the community from their beliefs. What was at stake was the belief system itself.

The Crusades

There are few more remarkable ideological expressions than the Crusades, a 200-year series of largely suicidal expeditions of more than half a million European men, women, and children to a destination halfway around the known world, in order to wrest control of holy places from a foreign and strange adversary that few of the expeditioners would have contacted without journeying for months over land and/or risking long and dangerous sea voyages. Historians have cited many factors as motivating the authorities and institutions that supported the Crusades and the people of all ranks who participated in them. These include efforts to consolidate Papal authority and otherwise increase order and discipline within the ranks of the Church, response to the pressures of increasing population, provision of aid to the Eastern Christian Community in Byzantium, and then the looting of that same city and others. The consensus among historians, however, is that these were in a large and important part holy wars whose aims were to secure control over and/or access to Jerusalem and its monuments central to the Christian faith of the Crusaders [e.g., 25–33]. "Religion permeated their existence to such an extent that it regulated their lives and shaped their individual and collective beings" [25, p. 17], and it was the "religious zeal that gripped even the fiercest warriors" that characterized the Crusades [27, p. 1]. Accounts written at the time of the Crusades support this view.

The first Crusade was called for by Pope Urban II on November 27, 1095 at the conclusion of the Council of Clermont. The Pope had indicated early in the council that he would make an important announcement at its conclusion, and so many thousands gathered for his final address that the location was moved to a large outdoor area surrounding a small hilltop. Upon his call for the Christians to take arms to protect and regain control of holy sites in Jerusalem, leader after leader, and ordinary citizens by the thousands, then and there committed themselves to the enterprise, and, in the language of the time, "took the cross." Several contemporary accounts of his speech have survived. According to Robert of Rheims, the Pope explained

that a people from the kingdom of the Persians, a foreign race, a race absolutely alien to God . . . has either completely razed the churches of God to the ground or enslaved them to the practice of its own rites. These men have destroyed the altars polluted by their foul practices . . . Jerusalem is the navel of the world. . . . This royal city, placed at the centre of the world, is now held captive by her enemies and is enslaved to pagan rites. . . . May you be especially moved by the Holy Sepulchre of our Lord and Savior, which is in the hands of unclean races, and by the Holy Places, which are now treated dishonourably and are polluted irreverently by their unclean practices. [from Robert of Rheims, *Historia Iherosolimitana*, RHC Oc., 111, pp. 727–730 quoted in 30, pp. 43–44].

Guilbert of Nogent reports that the Pope entreated those present:

You dearest brothers, must take the greatest pains to try to ensure that the holiness of the city [Jerusalem] and the glory of his Sepulchre will be cleansed, for the gentiles by their presence continually sully them insofar as they can. . . . Until now you have fought unjust wars . . . Now we are proposing that you should fight wars which contain the glorious reward of martyrdom. . . . [Guilbert of Nogent, *Historia quae dicitur gesta Dei por Francos*, RHC Oc. iv, pp. 137–140, quoted in 28, p. 46].

Pope Eugenius III initiated the second crusade with his papal bull "Quantum praedecessores" on December 1, 1145, in which he recalled that in regard to Urban II's exhortation, warriors from France and Italy "came together inflamed with the ardour of charity to liberate our Savior's glorious Sepulchre from the filth of the pagans" [quoted in 29, p. 121].

The Bishop of Freisingen, half-brother of Conrad III, the King of Germany, describes the widespread response in Germany to calls for the Second Crusade. Conrad, having already dedicated himself to leading a crusading army, traveled to Bavaria with the abbot of Ebach to gain additional recruits.

When the letters from Pope Eugenius and the abbot of Clairvaux had been read, he [the abbot of Ebrach] gave a brief homily, through which he persuaded nearly all those present to profess for the expedition. Having been roused already by rumours they had heard, all those present ran of their own accord to take up the cross . . . three bishops took up the cross . . . also Henry, duke of Bavaria, the king's brother, and countless men from the ranks of counts, nobles and knights. Even, strange to say, a crowd of brigands and thieves arrived, so great that no sane person could doubt that a change, as sudden as it was unexpected, had happened by the hand of God . . . Welf also, . . . one of the most noble princes of the realm, professed for the expedition with many followers on Christmas Eve . . . Vladislav, duke of the Bohemians, Ottokar, Margrave of Styria, and the illustrious Count Bernard of Carintha took up their crosses not much later, together with a large retinue [quoted in 29, p. 125].

The chaplain to Louis VII, King of France, describes a similar enthusiasm in his countrymen. After Louis VII decided to take up the cross, Bernard of Clairvaux came to give the Pope's blessing. "He mounted the platform, accompanied by the King, with a decorated cross. When he had spoken, everyone around began shouting for crosses" [quoted in 29, p. 124].

The kings of Germany and France themselves led armies on the Second Crusade, and the kings of Germany, France, and England led armies on the Third Crusade. Accounts and letters from the time indicate that the motivation of these kings was drawn from the exhortations of the popes. The chaplain to Louis VII explains that when Louis announced he would take up the cross, "zeal for the faith burned and shone in the king, together with contempt for worldly pleasure and temporal glory: he was in himself an example more persuasive than any speech" [quoted in 29, p. 122].

Henry II, king of England, wrote to beleagured Christians in the East, "According to what our sins have deserved, the Lord, in his divine judgment, has allowed the land which was redeemed by his own blood to be defiled by the hands of un-believers. I and my son Richard, among other princes, rejecting the trappings of this world and scorning all its pleasures, putting aside everything which is of the world will shortly, with the help of the Lord, visit you in prison with all our power" [quoted in 29, p. 165].

An ecclesiastic described the assembly at which Frederick 1, King of Germany took the cross for the Third Crusade. "Here, and not without the copious tears of many men, the Holy Roman Emperor Frederick I took up the sign of Christ's cross and declared that he was preparing himself for the memorable journey of Christ. . . . No one at that time in the whole of Germany was thought to be of manly standing who was seen without the sign of the life-giving cross. . . . In those boldest of combatants there burned the glorious passion for fighting against those who had invaded the Holy City and the Holy Sepulchre" [quoted in 29, p. 169].

In 1191, Richard the Lionheart, by then King of England, wrote from the campaign in the East to his judicial officer in England after he, together with the French army, had retaken the city of Acu. "We . . . are more concerned with the love and honour due to God than . . . acquisition even of more territories. Nevertheless, as soon as we have restored the territory of Syria to its original status [which was not subject to the king of England!] we shall return home" [quoted in 29, p. 189].

Although the crusading armies were formed in order to protect holy sites in and around Jerusalem, groups within Europe who, like the Muslims, did not share the Roman Catholic belief system, proved equally worthy of their attention. As Conrad III prepared the German army to set out for the east, several German princes received approval and encouragement to destroy or convert the Wends. Bernard wrote: "The devil . . . has aroused his depraved offspring, his wicked children, the pagans whom the fortitude of the Christians has endured far too long, disregarding to its own peril the pagan traps and treachery, rather than grinding down under its heel their poisonous heads . . . we announce that another Christian force should be armed against the pagans in order to destroy them utterly, or at least to convert the nations to take up the standard of salvation, promising to them and to those who have set out for Jerusalem the same indulgence for their sins. It was agreed by everyone in the council at Frankfurt that copies of this letter should be sent everywhere, and that the bishops and priests should read it out to the people of God, and that they should sign themselves with the sign of the holy cross and arm against the enemies of the cross of Christ who are beyond the Elbe" [quoted in 29, p. 127].

The Wends, a branch of the Slavic people who lived between the rivers Elbe and Saale and the Oder, had successfully resisted a 300-year effort by missionaries and soldiers to persuade or pressure them to abandon their beliefs and convert to Christianity. Indeed, "Helmut of Bosan, writing in the late 1160s, considered that the Slavs had remained the most obdurate of all the peoples of the north in the face of attempts to spread Christianity" [29, p. 128]. Although it could be argued that desire to possess their lands was an equal or greater motive than eradicating the belief system of the Wends, the same cannot be said for attacks by the Crusaders against Jews in Europe, since the latter did not occupy a region or own land.

In the early stages of both the first and second crusades there were multiple attacks on Jews throughout the Rhineland. Otto of Freisingen attributes the attacks against the Jews during the second crusade to the preaching by Radulf "that the Jews who lived in all these cities and towns should be slaughtered as if they were the enemies of the Christian religion. . . . this doctrine . . . took root so firmly that many Jews were killed in violent uprisings" [quoted in 29, p. 125]. A chronicler from the time reports "just one example from many of the slaughter of the Jews, which took place at Wurzburg in the month of February, 1147." A man's body was found chopped up into many pieces. "As if this gave them a just cause against the Jews, both the citizens and the pilgrims were seized by a sudden frenzy, and broke into the homes of the Jews, killing old men and young, women and little children without discrimination, without delay and without mercy" [quoted in 29, p. 126]. The fact that the violence of the Crusaders was directed against the Wends, the Jews, and the Muslims, all of whom have religious belief systems and customs that differ from those of the Crusaders but have little else in common, lends further support to the view that the Crusaders' stated goal to eliminate alternative belief systems was indeed an important motive for their action.

Accounts of Muslim observers and participants in the struggle with the Crusaders make it clear that they too understood the conflict to be, at least in large part, one of competing belief systems and religious practices. Usama Ibu Mungidh, a Muslim who had become accustomed to worshipping in a small oratory beside the Aqsa mosque in Jerusalem, which the Christians had made into a small church, describes an encounter with a pilgrim newly arrived from France that recalls the encounter between the ecotourist and the Chinese railroad cook described earlier.

One day I had risen to begin my prayers, when a Frank [the Muslim term for all Europeans] threw himself on me from behind, lifted me up and turned me so that I was facing east. 'That is the way to pray' he said. Some Templars . . . took him out of my way, while I resumed my prayer. But the moment they stopped watching him he seized me again and forced me to face east, repeating that this was the way to pray. Again the Templars intervened and . . . said: 'He is a foreigner who has just arrived today from his homeland in the north and he has never seen anyone praying facing any other direction than east.' [quoted in 29, p. 147]

Muslim views on the effect of Christians on Muslim holy places parallel the views of the Christians on the effects of Muslims on Christian holy places. Commenting on a visit to the city of Acre after it had fallen to the Christians, Ibn Jubayr writes:

May God exterminate [the Christians in] it and restore it to the Muslims. . . . The Franks ravished it from Muslim hands in 1104 and . . . its loss was one of Islam's griefs. Mosques became churches and minarets bell-towers, but God kept undefiled one part of the principal mosque . . . near the tomb of the prophet Salih . . . God protected this part [of the mosque] from desecration by the unbelievers for the benign influence of this holy tomb. [quoted in 29, p. 185]

Particularly revealing is the exchange of messages, quoted in the *Gesta Francorums*, between the Christian leaders of the first crusade and Kerbogha, a leader of Muslim forces, as the Crusaders sought to negotiate a truce. Envoys carried this message from the Christian leaders: "Our leaders and command-ers are amazed—utterly amazed—that you have entered so boldly and arrogantly the land that belongs to the Christians and to themselves. We suppose that perhaps you have come here because you wish at all costs to become Christians, or have you come here in order to afflict the Christians at all costs. All our leaders ask you to withdraw at once from the land of God and the Christians, which the blessed apostle Peter long ago converted by his preaching to the worship of Christ. But they allow you to take away all your belongings with you."

Kerbogha replied: "We do not choose or want your God or your Christianity, and we spit them out, and you too. . . . Return quickly and tell your leaders that if they wish at all costs to become Turks, and are willing to deny the God whom you worship on bended knee, and to cast aside your laws, we will give them this land and plenty more, and cities and castles in such abundance that none of your men will stay foot-soldiers, but will all be knights like us; and we will always consider them the greatest of friends. But if not, let them know that they shall pay the penalty of death, or . . . in perpetual captivity they will serve us and our children for ever" [quoted in 29, p. 83].

Both sides indicated clearly that what was most important was that the adversary abandon their belief system. Possession of land or capture of booty was a secondary concern at most. When the crusaders defeated Kerbogha and captured the city of Antioch, their leaders wrote to Pope Urban II that "our Lord Jesus Christ transferred the whole city of Antioch to the Roman religion and faith" [quoted in 29, p. 85]. The most important spoils of war brought back to Europe by victorious armies were religious relics—apostles' bones, stones, and dust from places mentioned in the Old and New Testaments, and, most important of all, wooden fragments and nails from the cross on which

Jesus had been crucified, as well as thorns from the crown of thorns and pieces of Christ's clothing. Elaborate chapels and reliquaries were built to house these objects [29, p. 166].

The huge cost of the Crusades in lives and resources provides another index of the importance attached to the conflict of belief systems. Indeed, with the cost of defeat and the rewards of victory measured in ideological terms, an accounting of worldly costs and balances provided only limited constraints. In the first Crusade, several armies set forth numbering in total close to 200,000 and varying widely in organization, preparation, and training. Most died along the way from illness or hostile armies. Among the 200,000 who began, there was a relatively well-trained and organized force of 80,000 men under the joint leadership of nobles from France and England. Twenty thousand survived 2 years of travel and battle, and in a fierce and bloody fight conquered Jerusalem. News of this success led to the dispatch of another 160,000 men from Europe. Most perished en route, and few or none reached the Holy Land.

After Pope Eugenius III exhorted the emperor of Germany and the king of France to defend the cross in the Second Crusade, another 140,000 men went to the East. A remnant of the force returned 2 years later without having had success. When the infidels retook Jerusalem in 1187, nearly 100 years after the First Crusade, the monarchs of the three principal European countries—Germany, France, and England—decided to personally lead their armies to recapture the city. The German monarch and his son died in the process, the French soon returned home, and the English fought for 2 years before concluding a treaty that ensured Christian pilgrims access to the Holy Sepulchre in Jerusalem.

Then, roughly a decade later, another Crusade followed that was so fantastical that its historic truth has at times been doubted. Apparently, a French peasant boy, encouraged by priests, proclaimed it his divine mission to lead a crusade of children to the Holy Land. Simultaneously, a boy raised a similar children's army in Germany. Together they led thousands of children, by land and sea, on a journey that for most ended in slavery or death. The final Crusade, led by Louis IX, monarch of France, had the explicit goal of converting the Moorish king of Tunis to Christianity. Louis IX and large numbers of his knights died, however, shortly after landing on the North African coast, and this unlikely enterprise quickly ended.

Cyprus and the Balkans

Conflict between Christian and Muslim peoples both preceded and persisted after the Crusades, most regularly and violently in the repeatedly contested contact zones between West and East in the Balkan states of Europe and on the Island of Cyprus. Here the situation was complicated further by the presence of Eastern Orthodox and Roman Catholic communities in close proximity. The anthropologist Jack Goody describes some of the differences between the Greek Orthodox and Muslims on Cyprus, differences that extend well beyond differences in theology and involve fundamental aspects of the relationship to the environment:

The two groups in contemporary Cyprus each have their own sacred and secular scripts, Greek in the first case, Arabic in the second, which ensures that they cannot read one another. But they also display strongly opposed ideologies, not simply relating to the obvious religious aspects of Christianity and Islam, but to all imagery of living things, a fact which sets them apart as far as objects in the world are concerned. And the world around them constantly reinforces this opposition which is much more omnipresent than the notion of ethnicity or even formal religion would suggest, leaving the opportunities for reconciliation thin and ineffective. [34, p. 113]

For an example of these central differences, Goody describes the manner in which the Orthodox worshippers and pilgrims bow their heads before and then kiss the religious icons that adorn their churches. "Here was not only the acceptance of figurative representations but also their worship or reverence as if they were living beings, or were the represented being themselves" [34, pp. 110–111]. What is represented is the inner nature, belief systems, and neurocognitive structures of the human creators of the icons, and it is the direct consonance of the internal and external that is celebrated and reassuring. This type of figural representation is an anathema to the Muslims, and the type of worship is considered both heathen and primitive.

When Christians and Muslims on Cyprus were not physically attacking one another, they attacked each other's symbols and places of worship. When the crusaders captured Cyprus in 1191, they built a fine gothic cathedral. When the Turks successfully invaded the island in 1570, they converted the cathedral into a mosque for their own worship, as they did with the cathedral built by Christians in Istanbul itself. All representations—paintings or sculpture—were removed. Sculpted features on the building itself that might be imitative of living things were hammered out. Here, then, is direct evidence of ridding the environment of prominent symbols of a different belief system.

The situation in the Balkans was further complicated by the presence of two large Christian communities that were as different from each other in world view as they were from the

Muslims. Catholicism, arising in the West, emphasizes ideas and deeds, and Catholic monks actively engage in worldly activities such as teaching nonreligious subjects, writing, and community service. Eastern Orthodoxy emphasizes beauty and magic, its religious services make even the most baroque Catholic services seem austere and intellectual, and Orthodox monks live contemplative lives [35]. Deep differences again extend into multiple aspects of life as a Catholic from Zagreb quoted by Robert Kaplan indicates: "When I entered the Yugoslav army, I met Serbs [Eastern Orthodox Christians] for the first time in my life. They told me that a traditional Serbian wedding lasts four days. Four days of prayers and feasting. Who needs that? One day is enough. After that you should go back to work. The Serbs struck me as weird, irrational, like Gypsies" [35, p. 25].

Division among the people of the Balkans is primarily on the basis of belief systems [35, 36]. The Roman Catholics, Greek Orthodox, and many Muslims have a common Slavic ethnicity. They speak the same language and often have the same names. Yet violence among the three communities has been repeated and horrific. Under Nazi occupation, Catholic Croatians are estimated to have murdered between 60,000 and 200,000 Orthodox Serbs (and thousands of Jews and Gypsies as well [35]). During the 1990s, tens of thousands more were killed in all communities and the violence only ended with outside intervention. Animosities were most intense among people who live in separate, religiously homogeneous mountain villages; apparently familiarity with one another over hundreds of years of even some contact in the large cities moderated the violence to a limited extent [35]. The perpetrators of intergroup violence identified themselves by explicitly religious symbols and ceremonies. Destruction of churches by the Muslims, and mosques by the Christians was widespread, and victims were made to sing the religious songs of the victors. Goody concludes that in Cyprus and the Balkans, in Ireland between the North and the South, in Europe among the Huguenots, Calvinists and Lutherans, in Islam between the Sunnis and Shiites, and in India among the Jains, Buddhists and Hindus, people who share the same language, race and other physical characteristics are divided and moved to violence by the ideology, world view, and practices and beliefs of religion.

Summary and Conclusion

For 80,000 to 100,000 years human beings lived in isolated communities distributed around the globe. Division into separate communities may have preceded the development of much of a language or culture, and there may never have been a common human language or culture as we think of each today. Certainly cultures developed independently of one another over most of the history of the species, and each community was unaware that most of the others even existed. The distinguishing feature of the current epoch in human development is the discovery and initiation of contact among previously separate and very different peoples and cultures.

In previous chapters, evidence was presented that demonstrated the importance of a close fit between internal neuropsychological structures created to conform with an individual's sensory and interpersonal environment at the time of development, and the environment in which the adult individual later finds him or herself. People are a particularly important aspect of that environment, both during and after development. Furthermore, human beings, to an extent unapproached by other animals, alter the environment in which their offspring develop, and children today are raised in an almost exclusively human-made cultural environment. In chapter 4 we saw the distress occasioned by major changes in the interpersonal environment through bereavement, and in the interpersonal and cultural environment through immigration.

This chapter has focused on the distress resulting from the meeting of previously separate and different cultures. The existence of very different peoples was widely imposed upon conscious awareness throughout the globe during the sixteenth, seventeenth, eighteenth, and nineteenth centuries. It proved an earth-shattering discovery that occupied the Euro-American imagination throughout the eighteenth and nineteenth centuries, as shown by the intense public interest in travel writings, attendance at lectures by explorers, exhibits of foreign peoples, and the activities of learned academic and scientific societies. The responses to exposure to foreign people and cultures were in many ways similar to the responses of individuals confronted with disparities between their personal belief systems and new environmental input within the context of their own general culture. Information about the foreign peoples and cultures was ignored, the people and cultures were devalued, and perceptions were distorted so that the new information could be incorporated into existing belief systems.

With continued contact between foreign cultures, these mechanisms proved insufficient to manage the distress caused by undeniable differences in behavior and beliefs, and each culture made efforts to get members of the other to act in ways that were consistent with the roles and qualities assigned them by the belief systems of the perceivers. Not infrequently, these attempts were followed by efforts to eliminate the continuingly offensive perception of difference through the seemingly paradoxical deployment of both military and missionary forces. Efforts to defend one's own belief system from the contradictory presence of another have led to some of the most fantastical extremes of behavior and violence, including for example the thousand-year war between Christian and Muslim communities in battles shifting across northern Africa and southern Europe and often targeting the holy places, religious icons, and religious leaders of the antagonist culture.

The emphasis on rationality that has dominated Euro-American intellectual thought since the seventeenth century has been loath to assign commitment to beliefs such as religion or ideology sufficient motivational force to account for major aspects of international discord and history. A reasoned economic self-interest or the inevitable developmental logic of nation-states and economic organizations have been seen as more reasonable explanations. Even ethnically inspired violence has been considered a more comfortable construct, perhaps because it is clothed in the scientism of biology that promises an eventual rational explanation. Indeed, it has maintained its appeal despite the absence of both a definition of the term *ethnic* and any explication of the mechanisms that might lead from the construct to the violence observed among peoples.

This book argues that differences in belief systems can themselves occasion intercultural violence, since concordance between internal structure and external reality is a fundamental human neurobiological imperative. This argument thus provides a rational basis for the apparent fact that people fight because of differences in religion and other beliefs; they fight to control the opportunity to create external structures that fit with their internal structures, and to prevent others from filling their environments with structures and stimulation that conflict with their internal structures.

The epilogue surveys contemporary conflicts among cultures and considers the special problems arising from the influence of one culture upon the children of another. Children inevitably become different than the representations of them that remain in the minds of their parents. In the contemporary context of global mixing of cultures, disjunctions between the actual child and the parent's representation of their child have been amplified, creating a powerful unease in adult communities. Culture itself has become a commodity, and a combined force of economics and ideology now drives its dissemination, making retreat from the intercultural contact zones impossible and battles for control of the cultural environment a common occurrence.

Epilogue

The United Nations estimates that 5,000–6,000 different languages are spoken today, each hundreds or thousands of years in evolution, each inseparably linked to a distinct culture, each the basis of the identity of a particular group of people, and each evidence of a group of people that was separate enough from others to develop its own language [this does not include dialects; see ref. 1]. Through the course of human history there have been still more languages, but some are no longer spoken. The United Nations' report goes on to state that there has been a dramatic increase in the rate of death and disappearance of languages over the past 200 years that is due largely to increased contact and ensuing competition among cultures. Based on fieldwork by experts, the report concludes that more than 3,000 languages spoken today are in danger of extinction. To each is linked the fate of a culture, the identity of a people, and the lives of thousands of individuals trying to hold on to the worlds that shaped them.

Threatened Indigenous Cultures

The Pennan of Malaysia are an example of a people with a dying culture and language [2]. The Pennan were nomadic hunters and gatherers displaced by companies that cut trees for export. All but 300 of the 7,000 Pennan people now live in government settlements with comfortable beds and zinc-roofed houses, but little productive work, little food, and no connection to the activities and surroundings that shaped them as a people and as individuals. The children watch television shows in a Malaysian language that few of them understand, but which most will probably speak by the time they are adults. An American anthropologist described a recent visit to the Pennan, his third; he was accompanied by a Canadian linguist making his tenth visit in an effort to compile the first Pennan grammar and dictionary. They were greeted warmly in the government settlement, but the Pennan men they knew from previous visits were embarrassed by how little food they could offer when it came time for meals.

At the request of the anthropologist, a small group of Pennan men led a 3-day excursion to visit one of the last bands of Pennan nomads living in a remote section of a national park that was off-limits to loggers. They found four families camping on a ridge where the Pennan have come for generations. Each morning began with a prayer: "Thank you for the sun rising, for the trees and the forest of abundance, the trees that were not made by man but by you." Toddlers and pet monkeys scampered about the camp, older children collected fruits and vegetables from the forest, adults made flour from the pith of sago palms, and a hunting party returned with two wild pigs. For the Pennan, everything one hears in the forest is an element of the spirit. They believe the plants and animals communicate with one another, and they obtain direction for their own behavior from the sounds of their forest world. Asik, the headman of this group of families, was temporarily jailed for obstructing logging in the forest. He explained his reaction to the changes in his world: "From the time of our origins we have preserved the trees and animals, every single thing in the forest. This we know. It is in our legends, our traditions. When we think of the places and our land, our hearts are troubled. Everywhere I go, I feel the need to weep" [2].

The world is dotted by indigenous peoples like the Pennan who are clinging desperately to vanishing traces of familiar habitat, or are displaced like fish out of water to foreign surroundings that are incapable of supporting their lives. The groups are small, 60,000 on average, but in aggregate the phenomenon is huge. The United Nations lists 5,000 of these indigenous cultures worldwide, totaling 300 million people [2]. The United Nations declared an International Day of the World's Indigenous People in August 1998, and a Decade of the World's Indigenous People, which we are in the midst of. These declarations are part of an effort to stop ethnocide, protect indigenous cultures and languages, and ensure the ability of indigenous peoples to live. The Guatamalan representative to the planning body declared that "when the principles and spirituality of indigenous people continued to be as real as in ancient times, in the relation between man and nature, it would be possible to say the Decade for the Protection and Promotion of Indigenous People would be on the right path" [3]. The representative from Mexico said, "there was an explosive phenomenon of collective identities demanding the right to be recognized, maintained and developed. In so doing they were seeking the transformation of the entire structure of society, challenging globalization . . . in the name of cultural uniqueness" [3]. In hoping to establish a middle ground, the United Nations Educational, Scientific and Cultural Organization seeks to promote the right to learn in one's own mother tongue, as well as the right of access to languages of wider communication.

Despite efforts like those of the United Nations, it is difficult to imagine that many indigenous peoples and cultures like the Pennan will continue to exist. Ours is an era marked by a dramatic and unprecedented reduction in the diversity of human cultures. Children in these disappearing cultures face the difficult challenges of learning new ways of life, dealing with dislike and distrust by members of the dominant culture in which they will live, and negotiating these, as well as the more ordinary challenges of life, without the usual support and guidance of parents and other older family members. Undoubtedly, many will succumb to substance abuse problems, and many others will have only limited opportunities for self-development and personal satisfaction.

More tragic still is the plight of the adult members of these disappearing cultures. Their worlds are being pulled out from under their feet and most are unable to learn and adjust to the ways of a new culture. Even when they make the effort, their neurobiological development is out of synchrony with the opportunities for skill and role development in the new culture. There is no match in their new environments for the knowledge and skills they developed in their original environments. The new schools that offer their children some hope for the future rob the parents of their most prized possessions and make clear the end of the world they knew. As an elderly midwife from the Ariaals, nomadic cattle herders from the Ndoto Mountains in Kenya, said: "We send our children to school and they forget everything. It's the worst thing that ever happened to our people" [2]. It can sometimes be difficult to believe that people from cultures very different from our own have feelings of hope, disappointment, fear, loss, and depression like ours. Or that they feel happiness, peace, and satisfaction in realizing childhood goals, contributing to and receiving the approval of a vital community to which they belong, and assuming the roles of the parents and other community elders they had admired. However, all evidence is that such feelings are universal, and so they must be if the neurocultural hypotheses developed in this book are accurate. And if they are, we are in the midst of an unprecedented global cultural transformation that is cutting hundreds of millions of human beings adrift from the essential sources of peace, happiness, and security, and denying them the opportunities to live their adult lives as fully functioning human beings.

Threatened National Cultures

Hundreds of large cultures also exist today safe from immediate threat of extinction and often secured by national boundaries. Even among these big players on the cultural scene, however, cultural conflict is evident and reflects the same issues that have played out to conclusion between small and large cultures. As the mass media of radio and television began to assume a commanding role in the distribution of information and ideas, and in shaping the thoughts and minds of people, governments of the free societies of North America and Europe acted to prevent foreigners from controlling their airways. In the United States, the Communication Act of 1934 banned aliens and foreign corporations from holding a broadcasting license, from owning more than 20% of an American company that owns a license, or from owning more than 25% of a company that controls a company with a license. Other North American and European countries passed similar laws or nationalized the major television and radio stations. These measures have proved insufficient because movies and television programs were offered for export and broadcast in countries other than those in which they were made.

The United States became the major supplier of such materials, and governments from other countries moved to protect their cultures, with efforts often directed by national ministries of culture. Even countries that allow United States military forces on their territory, that are partners with the United States in mutual defense treaties and join the United States in military action, are seeking to defend themselves against the encroachment of United States culture. The Canadian Broadcasting Act B-9.01, assented to on February 1, 1991, asserts that the Canadian broadcasting system "provides through its programming a public service essential to the maintenance and enhancement of national identity and cultural sovereignty" [section 3.1b]. It stipulates that "each broadcasting undertaking shall make maximum use, and in no case less than predominant use, of Canadian creative and other resources" [section 3.1f]. The purpose, stated in the legislative act, is to "encourage the development of Canadian expression by providing a wide range of programming that reflects Canadian attitudes, opinions, ideas, values and artistic creativity, . . . and by offering information and analysis . . . from a Canadian point of view" [section 3.1d]. As one commentator put it, "The people of Canada and the people of the United States are so thoroughly at peace with each other that relatively few issues ignite passions north and south of the famously undefended border. . . . The greatest heat currently surrounds Canada's defense of Canadian culture against the encroachment of U.S. influence and domination" [4, p.

203]. This defense includes use of subsidies, quotas, penalties, and prohibitions, and extends to broadcast and satellite television, film, music, books, and periodicals.

The European community put aside issues among its member countries to make its own effort at resisting the invasion of United States culture, issuing a directive requiring that at least half of television air time be dedicated to European-made programs. Of course, compliance with these measures is a different matter, and it is estimated that only 60–70% of stations meet their quota [5]. It is still another matter to influence what shows and movies viewers actually watch. In 1966 South Korea enacted legislation stipulating the number of days that Korean theaters must show Korean-made films. This quota now stands at 146 days a year, but between 1993 and 1998, only 15–25% of actual ticket sales were for showings of Korean-made films [6]. More recently, concern and protest in Europe has focused on the increasing presence of United States businesses that symbolize United States culture. McDonald's food places in France and Belgium were trashed or destroyed despite company statements that 80% of the products served in France are made in France and cooked by local employees. The French farmer who became a national celebrity after leading one attack explained that it is a "battle against globalization and for the right of people to feed themselves as they choose" [7]. A French political analyst explained that "Behind all this lies a rejection of cultural and culinary dispossession" [7].

The concern among the European military allies and economic trading partners of the United States extends even to fear for the safety of their languages. In 1994 the French legislature passed a law that French must be used in marketing all goods and services within France. Words in other languages may appear on products or in advertisements, but "any inscription or announcement made in another language must not, by virtue of its size, script, color or any other reason, be better understood than the French version" [8, p. 179]. Despite the fact that "France's youth, rapacious consumers of American exports, ridiculed the law" [8, p. 181], Germany and Switzerland enacted similar laws. Protests sprang up in Switzerland when a local school authority started teaching English to 7-year-olds and conducting arithmetic and discussion classes in English [9].

The marked increase in the use of English throughout Europe in recent decades makes the practical value of learning English undeniable, and 70% of respondents in a survey by the European Union agreed with the proposition that everyone should speak English, but nearly 70% also said that their own language needed to be protected [9]. A representative of the European Bureau for Lesser-Used Languages in Europe sees a clear danger to European languages if English were to be used in economic and educational activities. "If you start to eliminate the intellectual [language] community and the economic [language] community, you can eventually kill off a language" [9]. A newspaper editorial objecting to English being the first foreign language taught to young Swiss children went one step further and suggested that the school superintendent who instituted the policy "will go down in the history books as the gravedigger of the Swiss identity" [9].

Western Europe, Canada, and the United States share hundreds of years of cultural exchange. Both Canadian and United States cultures were heavily influenced by peoples who emigrated from western Europe. The great philosophers, novelists, playwrights, composers, and artists of western European cultures are familiar to Americans and have helped shape American thought and culture. All those cultures are predominantly Christian, have similar codes of law, and share the same written alphabet. Despite this, expansion of the United States culture stimulates the concerns and protectionist responses in Canada and Europe just described.

When it comes to the expansion of United States culture into Asia and the Middle East, the situation is different and more volatile. There has been much less exchange and mixing between the culture of the United States and those of these regions, and the differences between them are deep and broad. The culture of the United States is much more alien to peoples of Asia and the Middle East than it is to peoples of Europe, and represents more fundamental differences in worldview and behavior. As a result, the threat felt by the cultures of Asia and the Middle East is greater and the response more violent. In the Cambodia of Pol Pot, having spent any time in schools in the United States was reason for Cambodian nationals to be put to death [10]. In Vietnam, the number of activities legally considered offensive to Vietnam culture, and usually of Western origin, was recently increased from 200 to 750 [11]. As wealthy young men and women in the United Arab Emirates have started wearing clothes, dancing to music, and even riding motorcycles made in the United States, the government has acted to preserve the local culture. The chief of police of a seaside town is quoted in the *Christian Science Monitor* as explaining that "Western influence has eroded family values and weakened parental authority. Our police will step up efforts to maintain social values in keeping with Islamic and Arabic traditions" [12, p. 1]. Much more ominous, of course, are the calls for a holy war issued by some Muslim clerics to rid Islam of influences from the United States, and the violence that has followed from these declarations.

Parental Responses to the Influence of Foreign Cultures on Their Children

Teenagers and young adults are more likely than younger or older people to adopt some of the features of an intruding culture. Unlike their younger siblings, they are of an age to begin venturing out of the family and into contact with foreign elements in public spaces. Unlike older adults, their brains and values are still forming and are susceptible to shaping by the environment. Neurobiological research indicates that the human frontal lobes continue to actively develop until a person is 20–25 years old, and these are regions of the brain thought to be closely associated with values, morality, emotion, and other personality traits. It may be that, as Clifford Geertz has suggested, social processes have played a role in evolutionary selection of human characteristics, and that the delay in frontal lobe maturation has increased over generations because it affords greater ability to incorporate the growing collective wisdom and latest innovations [13]. In our time of mixing and interpenetration of cultures, late maturation of the frontal lobes increases the ability of young adults to incorporate features of a culture that is changed from without, and thereby widen the difference between themselves and their parents. It may also be that modern rearing practices and extended educational curricula that prevent individuals from assuming adult social roles as early as in past centuries have the additional effect of delaying activation and expression of the genes that complete brain development. Thus these social processes would extend the period of brain developmental plasticity, further allowing young adults to incorporate features of foreign cultures manifest in their local environments.

In chapter 3 I stressed the importance of parental figures as a source of stimulation for their children. The reverse is true as well. Laboratory studies of nonhuman mammals have documented that contact with infants can elicit maternal behavior and produce changes in the mothers' brains. Virgin female rats withdraw from the odors of rat pups [14], while new mothers are attracted to these same odors, and will even care for pups other than their own [15]. After 4–7 days of exposure to pups, virgin females develop maternal behavior and show the same changes in brain structure seen in new mothers [16, 17]. The total dependence of mammalian infants on parental care for safety and sustenance would seem to mandate a special interest in and attachment to the infant by the parent. Informal observation suggests that new parents are much more responsive in a rapid and automatic way to infant distress cries than are their still childless peers. Psychoanalytic observations suggest that pregnancy is a period of profound physiological and psychological change that culminates in a reconfiguration of the self as forever part of a mother-child dyad [18]. It appears, then, that at the time that adults have children, enhanced neuroplasticity facilitates the incorporation of the offspring into the external world and in the internal neuropsychological structures of the parents. Even without such plasticity, of course, mere extended and intensive contact between parents and children would establish an important internal representation of the child in the parent, just as spouses incorporate representations of one another and suffer bereavement at the loss of a spouse, as discussed in chapter 4.

Whatever the developmental mechanisms, children are of special importance in the inner worlds of their parents, and therefore the match between the internal representation and the external reality of their children is of special importance to parents. This again is a roundabout way to get to something that is obvious from the start. People are generally saddened by the death of any child, but their grief, even when the child is that of a close friend, is nothing in comparison with the grief of losing one's own child. Their pleasure when a child performs well on a school exam is great if the child is their own and minimal if it is not. Their upset when a child dresses oddly, dyes their hair unnatural colors, and collects tattoos and body piercings is minor if the child is not their own and is major if it is their child (if of course they are a typical parent in the United States).

This powerful identification between internal structures of the parental psyche and external realities of the characteristics and accomplishments of their children can lead to intense feelings when the children no longer match their expectations. The behavior of parents at children's sports events often provides examples: fury at a child for performing below expectation; or at other children, coaches, referees, or other parents perceived as preventing a child from performing well enough. Odd as it seems to those not caught up in such feelings, this fury can lead to physical attacks on coaches, officials, or other parents [19]. These behaviors are so pervasive that twelve state legislatures in the United States have made it a crime punishable with imprisonment and/or fines to lay a hand on an official at children's sports games, and similar laws were before other state legislatures at the time of this writing [20]. Some children's sports leagues require parents to watch instructional videos and sign codes of behavior, while others require that spectators watch in total silence [19, 20]. The National Association of Sports Officials now offers assault insurance and legal assistance to its members.

The special importance of children to their parents, and the propensity for children to adopt aspects of an intruding culture, make the impact of one culture on the children of another a powerful cause of animosity between cultures. The aforementioned efforts to limit foreign movies and television shows, to boycott or destroy McDonald's and Disneyworld, and to delay and subordinate the teaching of English are in part efforts to protect children from the influence of a foreign culture. Perhaps the greatest threat is that one's child will marry someone from another culture, and then abandon, or compromise, a wide range of parental expectations. The following three examples illustrate the depth and ubiquity of this concern.

In 1816 the people in the small Connecticut town of Cornwall created a school for Heathen Youth. Prospective students were defined by their otherness rather than by their specific characteristics, and included Hawaiians, Greeks, and Native Americans. The goal of educating these young men and women in the customs and values of the town was widely and enthusiastically supported by the residents of Cornwall who, the historical record shows, made hundreds of contributions to support the school. When, however, two Native American male students married two town women, the school was promptly closed [21]. The United Arab Emirates has created a $150 million marriage fund to encourage young men to marry local women. The head of the marriage fund explained that "marrying within our own culture is much better for our society and for future generations. Children with non-Arab mothers will grow up confused about which culture they belong to. We want to avoid that" [12, p. 1]. The octogenarian president of a local Harvard University alumni club recently stated that he did not want his granddaughters to attend Harvard because the ethnic diversity of the student body created too great a risk that they would meet and marry young men from a different culture. Parental discomfort with the changes in a child that follow from an intercultural marriage are so great that in many cultures there is no further contact with the child, and in some the child is mourned as if he or she had died.

Role of the United States

Most of the violent conflicts raging around the world today are between peoples of different cultures and belief systems: Protestants versus Catholics in Northern Ireland; Muslims versus Hindus in India and Pakistan; Buddhists versus Hindus in Sri Lanka; Jews versus Muslims in the Middle East; and Christians versus Muslims in Chechnya, Indonesia, and Nigeria, for example. United States culture is locked in a struggle with other cultures, but the nature of the struggles differs in important ways from the more overtly religious conflicts between geographic neighbors. The United States is a largely Christian nation, but the scientific, business, and popular culture it exports is without explicit religious content. That culture appeals particularly to the youth in foreign cultures who are seeking education, business, and employment opportunities and are consuming popular culture. While it does not explicitly focus on religion, it depicts and affects behaviors that are of religious significance in non-Christian societies where religion plays a large and explicit role in social, legal, and governmental life. The legal separation of church and state in the United States, and the absence of a large explicit contribution of religion to popular culture in the United States, can make it difficult to appreciate the nature and intensity of the animosity to United States culture in non-Christian religious states.

The intercultural conflicts in which the United States participates also differ from others in not being limited to geographic neighbors with whom there is direct interpersonal contact. The culture of the United States is disseminated throughout the world through medical and scientific literature, mass media, and consumer products of popular culture, and the conflicts in which the country is embroiled are similarly widespread. The economy of the United States is the largest in the world, and its extension throughout the world spreads the English language and United States culture. Moreover, the economy of the United States has come to depend upon foreign resources, both material and human, and upon foreign markets to such an extent that it is not possible to limit the penetration of United States culture into foreign domains. The culture of the United States culture is a juggernaut barreling into conflict with its cultural first cousins in western Europe and North America and with the deeply alien cultures of Asia and the Middle East.

The United States also differs from other societies in the wide range of internal cultural struggles and accommodations that have marked its history. American soil is all too familiar with violence among peoples, with centuries of hostilities among Native American cultures, the virtual elimination of these cultures by invading Europeans, subordination of African-Americans as second-class human beings, a bloody civil war, exploitation of waves of immigrants that entered the country through "that terrible little Ellis Island," and internment of loyal Americans during World War II because they were of Japanese origin. It is also a soil on which people of different cultures and races have mixed and shared governance and wealth to a degree with few precedents. Both Catholics and Protestants have served as president, the narrowly defeated (or victorious) candidates for president and vice-president from the Democratic party in 2000 were a Protestant and a Jew, and leading government posts at all levels include individuals from many ethnic, racial, and cultural backgrounds. The secretary of state from 2000 to 2004 was an African-American man, something that the prime minister of Great Britain said at the time would not be possible in his country.

These successes at overcoming cultural conflict rest on common adherence to an overarching national identity with an explicit ideological commitment to the equality of people of all origins. Laws have been created to safeguard individual rights regardless of one's ethnic origins or cultural practices, and jurists have extensively debated the proper balance between the protection of free speech and the restriction of hate speech. Nonetheless, the existence of special divisions within many law enforcement bodies to deal with hate crimes speaks to the continued tensions among subcultures. And now, even within the borders of this self-conceived cultural melting pot, the global movement to preserve indigenous cultures is manifest in a rising assertion of a multiculturalism that strives to maintain subcultural ethnic identities within the general society. The United States is at once the primary cultural aggressor on the global scene and a country with great experience and success in living with and resolving cultural conflicts within its borders.

What Lies Ahead?

It is difficult to imagine that the violent cultural conflicts around the world will soon play out in a peaceful way. Instead, with the geographic barriers that allowed our species to develop thousands of distinct cultures over most of its 150,000–200,000-year history now eliminated, the bloody business of resolving differences in belief systems seems destined to continue along three types of battle lines for several more generations at least. One set of battles will continue between pairs of opposing cultures that live side by side.

A second set is developing as a result of the large numbers of Muslims from North Africa who, for demographic and economic reasons, are rapidly moving into what have been relatively homogeneous, Christian communities in Europe. The small French city of Dreux is an example [22]. Its population grew by only 50 people per year from 1801 to 1900 and only 88 per year from 1900 to 1950, but by 1,000 per year from 1954 to 1968. By 1970, 11% of the population were recent immigrants, most of whom were Muslims from Algeria. Only a year later, 16% of the population was foreign born. Over the next decade, ultra-right-wing political candidates with fascist leanings made unprecedented gains in local elections, promising to do all in their power to close the newly opened mosque, distributing political pamphlets that referred to the invasion of France by hordes of immigrants, and asserting that "the people of Dreux will defend their historical and cultural identity . . . the flow of immigration must be reversed" [22, p. 124]. In Holland, 10% percent of the population now consists of first-generation immigrants, many of them Muslims from Turkey and Morocco [23]. Most live in the major cities, and if the current population changes continue, the main Dutch cities will all have Muslim majorities by 2015. A liberal democratic society that long prided itself on openness and tolerance has rapidly found itself in an identity struggle. The issues came dramatically to national and international attention when a young Moroccan murdered a well-known Dutch filmmaker in broad daylight on a public street because the filmmaker had made an 11-minute film seen by some as disrespectful to Islam [23].

The third set of intercultural conflicts will be between the expansionist culture of the United States as it reaches global proportions and all other major cultures. Much of the way for this expansion has been cleared by the European colonial powers that destroyed indigenous cultures in large parts of the world and in others usurped political and military power and began the process of introducing Euro-American culture through missionary churches and English-language schools. In other parts of the world, however, large indigenous cultures have existed for millennia unaltered by colonization. The Islamic nations have fiercely and successfully contested the combat zone with the Christian nations of Europe for more than a thousand years, maintaining their distinct culture throughout this period. All evidence indicates their response will be the same when they are confronted by the culture of the United States. The ancient cultures of Asia await fuller contact with that of the United States, but the histories of violence among them suggest a similar resistance to cultural encroachment from the United States. Perhaps great statesmen or women will emerge who are capable of minimizing the bloodshed as the cultural confrontations proceed. Perhaps the successes at cross-cultural integration within the United States will somehow prove helpful on the international stage.

What lies beyond the violence? Efforts like those of the United Nations to preserve numerous small cultures, each with their distinctive traditional relation to their environment, are unlikely to succeed in any vital, human way. The environment, cultural and otherwise, is changing too substantially and regular contact with other peoples is part of that change. Many young people from these cultures will have contact with the broader, new culture and change because of it. They in turn will change the traditional culture if they remain within it. Of course, if members of dominant and expanding cultures around the small cultures could see these other people as equally human, and allow the smaller culture to change and fold into the dominant culture over several generations as do immigrant families in the United States, the human cost would be much reduced.

A multiculturalism within a broader national culture in which the individual cultural entities are maintained in stable traditional form is equally improbable. The youth of each cultural subgroup change too deeply from contact with the other cultures, and with the unifying national culture, to maintain their culture of origin in traditional form. Elements of many different cultures may survive as vital parts of a new culture, but they will be mixed with elements from other cultures in individual lives, as Jewish Americans may practice yoga derived from India, cook Chinese food, and enjoy writing Japanese haiku. Few individuals in the United States will follow a life that is in accord with integrated components from a single cultural source. So too will probably be the case on a global scale. Economic, scientific, medical, and entertainment interests will not leave the cultural behemoths of Euro-America, the Middle East, and Asia living side by side without contact. The angry consternation of their elders will not stop the youth of each culture from assuming characteristics from the others and then changing their cultures from within as they themselves assume leadership roles and act to make the external world consonant with their hybrid selves.

What then will happen when information flows over the geographic obstacles and cultural barriers that have divided human beings into thousands of isolated groups for most of our history as a species? Science fiction writers describe future cultures that are bleak in their homogeneity, filled with individuals whose thoughts and feelings are so much the same that they seem hardly alive. However, if this were the inevitable outcome of an integrated global culture, we would have to assign a similar existential fate to all individuals who passed their lives in the cultures of old that were each worlds unto themselves, unaware of or without contact with other cultures. It seems to me that such bleak predictions arise from discomfort at the prospect of losing the world of cultural diversity that has become familiar during the past 300 years. Our world should be as it is now, full of cultural variation and conflict, and anything else would be lifeless and dead.

One need only turn to the leading universities of the Western world to see an alternative vision of a global culture. Here large numbers of youths are given free access to information from and about all quarters of the globe and all spheres of human interest and knowledge. Outside of the classrooms and libraries, they choose from a panoply of social, athletic, employment, service, and artistic activities in which to engage, each itself a universe of more specific choices. An endless variety of niches are created and filled. An individual's activities and interests are influenced by their personal backgrounds, but the many antecedent experiences that influence these choices defy orderly prediction and add diversity even within small niches. The overall faculty-student community often feels itself divided into cultures with interests, priorities, and ways of thought so different from one another that special efforts are made to bridge the gaps among them and promote the exchange of ideas. Diversity within the university lies in the multiple, dynamic aggregates of lives that are similarly defined from the large and changing array of possible activities.

In the same way, a global culture can generate and sustain new diversities even as it transcends and erases the distinctions developed by people who lived for thousands of years in isolated communities. The cultural diversity we know today, a mixing of societies that were previously differentiated during extended isolation, would be replaced by a diversity born of contact among huge numbers of individuals shaped into fluid groups by the choices they make from a superabundance of educational and other activities. While these communities of thought, knowledge, activity, and customs might differ from ethnic and religious communities in interesting ways, I would not be the first to call them cultures [24], and they would create a richly heterogeneous human landscape. This to me is a more likely and appealing image for the future than those in Ray Bradbury's *Fahrenheit 451* or George Orwell's *1984*, but what else would you expect from a university professor?

References

Introduction

1. Hundert EM. *Philosophy, Psychiatry and Neuroscience.* Clarendon Press, Oxford, 1989.

2. Geertz C. The growth of culture and the evolution of mind. In *Theories of the Mind.* ed. J. Scher. Free Press of Glencoe, Macmillan, New York, 1962.

3. Geertz C. *The Interpretation of Cultures.* Basic Books, New York, 1973.

4. Lewontin RC, Rose S, Kamin LJ. *Not in Our Genes.* Pantheon Books, New York, 1984.

5. Cole M. *Cultural Psychology.* Belknap Press of Harvard University, Cambridge, Mass., 1975.

6. Horowitz DL. *Ethnic Groups in Conflict.* University of California Press, Berkeley, 1985.

7. Todorov T. *The Conquest of America: The Question of the Other.* Trans. R Howard, Harper Collins, New York, 1984.

8. Volf M. *Exclusion and Embrace.* Abingdon Press, Nashville, Tenn., 1996.

9. Goody JR. *Islam in Europe.* Polity Press, Cambridge, UK, 2004.

10. Pratt ML. *Imperial Eyes: Travel Writing and Transculturation.* Routledge, London, 1992.

11. Huntington SP. The clash of civilizations? *Foreign Affairs* 72: 22–49, 1993.

12. Diamond J. *Collapse: How Societies Choose to Fail or Succeed.* Viking, New York, 2005.

Chapter 1

1. John ER. Switchboard versus statistical theories of learning and memory. Coherent patterns of neural activity reflect the release of memories and may mediate subjective experience. *Science* 177: 850–864, 1972.

2. Chapin JK, Nicolelis MK. Principal components analysis of neuronal ensemble activity reveals multidimensional somatosensory representations. *J. Neuroscience Methods* 94: 121–140, 1999.

3. Gerstein GL. Analysis of firing patterns in single neurons. *Science* 131: 1811–1812, 1960.

4. John ER, Tang Y, Brill AB, Young R, Ono K. Double-labeled metabolic maps of memory. *Science* 233: 1167–1175, 1986.

5. Xu F, Kida I, Hyder F, Shulman RG. Assessment and discrimination of odor stimuli in rat olfactory bulb by dynamic functional MRI. *Proceedings of the National Academy of Sciences of the United States of America* 97: 10601–10606, 2000.

6. Kristan WB Jr, Shaw BK. Population coding and behavioral choice. *Current Opinion in Neurobiology* 7: 826–831, 1997.

7. Luria AR. *Human Brain and Psychological Processes.* Trans. B. Haigh. Harper and Row, New York, 1966.

8. van Melchner L, Pallas SL, Sur M. Visual behavior mediated by retinal projections directed to the auditory pathway. *Nature* 404: 871–876, 2000.

9. Hebb DO. *The Organization of Behavior.* Wiley, New York, 1949.

10. Kandel ER. Cellular mechanisms of learning and the biological basis of individuality. In *Principles of Neuroscience.* eds. ER Kandel, JH Schwartz, TM Jessel. McGraw-Hill, New York, 2000.

11. Kuhlenbeck H. *Vorlesungen uber das Zentralnervensystem der Wirbeltiere. Fischer: Jena, 1927.* [Kuhlenbeck H. *The Human Brain and Its Universe.*] Karger, Basel, Switzerland, 1982.

12. MacLean PD. *The Triune Brain in Evolution: Role in Paleocerebral Functions.* Plenum, New York, 1990.

13. Stephan H. Quantitative investigations on visual structures in primate brains. In *Proceedings of Second International Congress of Primatology, Vol. 3, Neurology, Physiology, and Infectious Diseases.* ed. HO Hofer. Karger, Basel, Switzerland, 1968, pp. 34–42.

14. Ploog D, Melnechuk T. *Neurosciences Research Symposium.* vol. 6, 1970.

15. Bownds MD. *The Biology of Mind.* Fitzgerald Science Press, Bethesda, Md., 1999, p. 77.

16. Corballis MC. Phylogeny from apes to humans. In *The Descent of Mind.* eds. MC Corballis, SEG Lea. Oxford University Press, New York, 1999.

17. Ekman P, Oster H. Facial expressions of emotion. *Annual Review of Psychology* 30: 527–554, 1979.

18. Darwin C. *The Expression of the Emotions in Man and Animals.* John Murray, London, 1872.

19. Charlesworth WR, Kreutzer MA. Facial expressions of infants and children. In *Darwin and Facial Expression.* ed. P. Ekman. Academic Press, New York, 1973.

20. DeRivera J. A structural theory of the emotions. In *Psychological Issues.* vol. 10. International Universities Press, Madison, Conn., 1977.

21. Fridja NH. *The Emotions.* Cambridge University Press, Cambridge, 1986.

22. Phan KL, Wager T, Taylor SF, Liberzon I. Functional neuroanatomy of emotion: A meta-analysis of emotion activation studies in PET and fMRI. *NeuroImage* 16: 331–348, 2002.

23. Dimberg V. Facial reactions to facial expressions. *Psychophysiology* 19: 643–647, 1982.

24. Ekman P, Levenson RW, Friesen WV. Autonomic nervous system activity distinguishes among emotions. *Science* 221: 1208–1210, 1983.

25. Lanzetta JT, Cartwright-Smith J, Kleck R. Effects of nonverbal dissimulation on emotional experience and autonomic arousal. *J. Personality and Social Psychology* 53: 354–370, 1976.

26. Laird JD. Self-attribution of emotion: The effects of expressive behavior on the quality of emotional experience. *J. Personality and Social Psychology* 29: 475–486, 1974.

27. Adolphs R. Neural systems for recognizing emotion. *Current Opinion in Neurobiology* 12: 169–177, 2002.

28. Carr L, Iacoboni M, Dubeau MC, Mazziotta JC, Lenzi GL. Neural mechanisms of empathy in humans: A relay from neural systems for imitation to limbic areas. *Proceedings of the National Academy of Sciences of the United States of America* 100: 5497–5502, 2003.

29. Kawasaki H, Adolphs R, Kaufman O, Damasio AR, Damasio M, Granner M, Bakken H, Hori T, Howard MA. Single neuron responses to emotional visual stimuli recorded in human ventral prefrontal cortex. *Nature Neuroscience* 4: 15–16, 2001.

30. Whalen PJ, Rauch SL, Etcoff NL, McInerney SC, Lee MB, Jenike MA. Masked presentations of emotional facial expressions modulate amygdala activity without explicit knowledge. *J. Neuroscience* 18: 411–418, 1998.

31. Malatesta CZ, Izard CE. The ontogenesis of human social signals: From biological imperative to symbol utilization. In *The Psychobiology*

of Affective Development. eds. NA Fox, RJ Davidson. Erlbaum Associates, Hillsdale, N.J., 1984.

Chapter 2

1. Rasch E, Swift H, Riesen AH, Chow KL. Altered structure and composition of retinal cells in dark-reared animals. *Experimental Cell Research* 25: 348–363, 1961.

2. Liang H, Crewther DP, Crewther SG, Barila AM. A role for photoreceptor outer segments in the induction of deprivation myopia. *Vision Research* 35(9): 1217–1225, 1995.

3. Kupfer C, Palmer P. Lateral geniculate nucleus: Histological and cytochemical changes following afferent denervation and visual deprivation. *Experimental Neurology* 9: 400–409, 1964.

4. Wiesel TN, Hubel DH. Effects of visual deprivation on morphology and physiology of cells in the cat's lateral geniculate body. *J Neurophysiology* 26: 978–993, 1963.

5. Hubel DH, Wiesel TN. The period of susceptibility to the physiological effects of unilateral eye closure in kittens. *J. Physiology* 206: 419–436, 1970.

6. Sherman SM, Hoffman KP, Stone J. Loss of a specific cell type from dorsal lateral geniculate nucleus in visually deprived cats. *J. Neurophysiology* 35: 532–541, 1972.

7. Sherman SM, Sanderson KJ. Binocular interaction on cells of the dorsal lateral geniculate nucleus of visually deprived cats. *Brain Research* 37: 126–131, 1972.

8. Hubel DH. *Eye, Brain and Vision*. Scientific American Library, W. H. Freeman, New York, 1988, chap. 9.

9. Tigges M, Tigges J. Parvalbumin immunoreactivity in the lateral geniculate nucleus of rhesus monkeys raised under monocular and binocular deprivation conditions. *Visual Neuroscience* 10(6): 1043–1053, 1993.

10. Aghajanian GK, Bloom FE. The formation of synaptic junctions in developing rat brain: A quantitative electron microscopic study. *Brain Research* 6: 716–727, 1967.

11. Cragg BG. What is the signal for chromatolysis? *Brain Research* 23: 1–21, 1970.

12. Fifková E. Changes of axosomatic synapses in the visual cortex of monocularly deprived rats. *J. Neurobiology* 2: 61–71, 1970.

13. Kumar A, Schliebs R. Postnatal laminar development of cholinergic receptors, protein kinase C and dihydropyridine-sensitive calcium antagonist binding in rat visual cortex. Effect of visual deprivation. *International Journal of Developmental Neuroscience* 10(6): 491–504, 1992.

14. Kumar A, Schliebs R. Postnatal ontogeny of GABA$_A$ and benzodiazepine receptors in individual layers of rat visual cortex and the effect of visual deprivation. *Neurochemistry International* 23(2): 99–106, 1993.

15. Robner S, Kumar A, Kues W, Witzemann V, Schliebs R. Differential laminar expression of AMPA receptor genes in the developing rat visual cortex using *in situ* hybridization histochemistry. Effect of visual deprivation. *International Journal of Developmental Neuroscience* 11(4): 411–424, 1993.

16. Rakic P, Suner I, Williams RW. A novel cytoarchitectonic area induced experimentally within the primate visual cortex. *Proceedings of the National Academy of Sciences of the United States of America* 88(6): 2083–2087, 1991.

17. Benson TE, Ryugo DK, Hinds JW. Effects of sensory deprivation on the developing mouse olfactory system: A light and electron microscopic, morphometric analysis. *J. Neuroscience* 4(3): 638–653, 1984.

18. Skeen LC, Due BR, Douglas FE. Neonatal sensory deprivation reduces tufted cell number in mouse olfactory bulbs. *Neuroscience Letters* 63: 5–10, 1986.

19. Najbauer J, Leon M. Olfactory experience modulated apoptosis in the developing olfactory bulb. *Brain Research* 674 (2): 245–251, 1995.

20. Berardi N, Cattaneo A, Cellerino A, Domenici L, Fagiolini M, Maffei L, Pizzorusso T. Monoclonal antibodies to nerve growth factor (NGF) affect the postnatal development of the rat geniculocortical system. *J. Physiology (London)* 452: 293P, 1992.

21. Berardi N, Domenici L, Parisi V, Pizzorusso T, Cellerino A, Maffei L. Monocular deprivation effects in the rat visual cortex and lateral geniculate nucleus are prevented by nerve growth factor (NGF). I. Visual cortex. *Proceedings of the Royal Society of London, Ser. B.* B251: 17–23. 1993.

22. Domenici L, Cellerino A, Maffei L. Monocular deprivation effects in the rat visual cortex and lateral geniculate nucleus are prevented by nerve growth factor (NGF). II. Lateral geniculate nucleus. *Proceedings of the Royal Society of London, Ser. B.* B251: 25–31, 1993.

23. Carmignoto G, Canella R, Candeo P, Comelli MC, Maffei L. Effects of nerve growth factor on neuronal plasticity of the kitten visual cortex. *J. Physiology (London)* 464: 343–360, 1993.

24. Pizzorusso T, Fagiolini M, Fabris M, Ferrari G, Maffei L. Schwann cells transplanted in the lateral ventricles prevent the functional and anatomical effects of monocular deprivation in the rat. *Proceedings of the National Academy of Sciences of the United States of America* 91(7): 2572–2576, 1994.

25. Kossut M. Effects of sensory denervation and deprivation on a single cortical vibrissal column studied with 2-deoxyglucose. *Physiologia Bohemoslovaca* 34: 70–83, 1985.

26. Melzer P, Crane AM, Smith CB. Mouse barrel corks functionally compensate for deprivation produced by neonatal lesion of whisker follicules. *European Journal of Neuroscience* 5: 1635–1652, 1993.

27. Kossut M. Effects of sensory deprivation upon a single cortical vibrissal column: A 2DG study. *Experimental Brain Research* 90: 639–642, 1992.

28. Kaas JH, Merzenich MM, Killackey HP. The reorganization of somatosensory cortex following peripheral nerve damage in adult and developing mammals. *Annual Review of Neuroscience* 6: 325–356, 1984.

29. Toldi J, Rojik I, Feher O. Neonatal monocular enucleation-induced cross-modal effects observed in the cortex of adult rat. *Neuroscience* 62(1): 105–114, 1994.

30. Cynader M, Berman N, Hein A. Cats reared in stroboscopic illumination: Effects on receptive fields in visual cortex. *Proceedings of the National Academy of Sciences of the United States of America* 70(5): 1353–1354, 1973.

31. Cynader M, Chernenko G. Abolition of direction selectivity in the visual cortex of the cat. *Science* 193(4252): 504–505, 1976.

32. Tretter F, Cynader M, Singer W. Modification of direction selectivity of neurons in the visual cortex of kittens. *Brain Research* 84(1): 143–149, 1975.

33. Blakemore C, Cooper GF. Development of the brain depends on visual experience. *Nature* 228: 477–478, 1970.

34. Hirsch HB, Spinelli D. Visual experience modifies distribution of horizontally and vertically oriented receptive fields in cats. *Science* 168: 869–871, 1970.

35. Callaway EM, Katz LC. Effects of binocular deprivation on the development of clustered horizontal connections in cat striate cortex. *Proceedings of the National Academy of Sciences of the United States of America* 88: 745–749, 1991.

36. Guthrie KM, Wilson DA, Leon M. Early unilateral deprivation modifies olfactory bulb function. *J. Neuroscience* 10(10): 3402–3412, 1990.

37. Wolf A. The dynamics of the selective inhibition of specific functions in neurosis: A preliminary report. *Psychosomatic Medicine* 5: 27–38, 1943.

38. Tuso RJ, Repko MX, Smith CB, Herdman SJ. Early visual deprivation results in persistent strabismus and nystagmus in monkeys. *Investigative Ophthalmology and Visual Science* 32: 134–141, 1991.

39. Murphy KM, Mitchell DE. Vernier acuity of normal and visually deprived cats. *Vision Research* 31(2): 253–266, 1991.

40. Tees RC. Effects of early auditory restriction on adult pattern discrimination. *J. Comparative and Physiological Psychology* 63: 389–393, 1967.

41. Tees RC, Symons LA. Intersensory coordination and the effects of early sensory deprivation. *Developmental Psychobiology* 20(5): 497–507, 1987.

42. Tees RC, Midgley G. Extent of recovery of function after early sensory deprivation in the rat. *J. Comparative and Physiological Psychology* 92(4): 768–777, 1978.

43. Wilson PD, Riesen AH. Visual development in rhesus monkeys neonatally deprived of patterned light. *J. Comparative and Physiological Psychology* 61: 87–95, 1966.

44. Tees RC. Effects of early restriction on later form discrimination in the rat. *Canadian Journal of Psychology* 22: 294–301, 1968.

45. Siegel J, Coleman P, Riesen AH. Deficient pattern-evoked responses in pattern-deprived cats. *Electroencephalography and Clinical Neurophysiology* 35: 569–573, 1973.

46. Batkin S, Groth H, Watson JR, Ansberry M. Effects of auditory deprivation on the development of auditory sensitivity in albino rats. *Electroencephalography and Clinical Neurophysiology* 28: 351–359, 1970.

47. Wilson DA, Wood JG. Functional consequences of unilateral olfactory deprivation: Time-course and age sensitivity. *Neuroscience* 49(1): 183–192, 1992.

48. Kandel ER. Cellular mechanisms of learning and the biological basis of individuality. In *Principles of Neuroscience*. ed. ER Kandel, JH Schwartz, TM Jessell. McGraw-Hill, New York, 2000.

49. Martin SJ, Grimwood PD, Morris RGM. Synaptic plasticity and memory. *Annual Reviews of Neuroscience* 23: 649–711, 2000.

50. Dudai Y. Molecular basis of long-term memories. *Current Opinion in Neurobiology* 12: 211–216, 2002.

51. Rosenzweig MR. Environmental complexity, cerebral change, and behavior. *American Psychologist* 21: 321–332, 1966.

52. Mirmiran M, Uylings HB. The environmental enrichment effect upon cortical growth is neutralized by concomitant pharmacological suppression of active sleep in female rats. *Brain Research* 261(2): 331–334, 1983.

53. Hyden H, Ronnback L. Incorporation of amino acids into protein in different brain areas of rat, subjected to enriched and restricted environment. *J. Neurological Sciences* 34(3): 415–421, 1977.

54. Altschuler RA. Morphometry of the effect of increased experience and training on synaptic density in area CA3 of the rat hippocampus. *J. Histochemistry Cytochemistry* 27(11): 1548–1550, 1979.

55. Bryan GK, Riesen AH. Deprived somatosensory-motor experience in stumptailed monkey neocortex: Dendritic spine density and dendritic branching of layer IIIB pyramidal cells *J. Comparative Neurology* 286(2): 208–217, 1989. Published erratum appears in *J. Comparative Neurology* 1989 Nov 22: 289(4): 709.

56. Held JM, Gordon J, Gentile AM. Environmental influences on locomotor recovery following cortical lesions in rats. *Behavioral Neuroscience* 99(4): 678–690, 1985.

57. Whishaw IQ, Sutherland RJ, Kolb B, Becker JB. Effects of neonatal forebrain noradrenaline depletion on recovery from brain damage: Performance on a spatial navigation task as a function of age of surgery and postsurgical housing. *Behavioral and Neural Biology* 46(3): 285–307, 1986.

58. Gentile AM, Beheshti Z, Held JM. Enrichment versus exercise effects on motor impairments following cortical removals in rats. *Behavioral and Neural Biology* 47(3): 321–332, 1987.

59. Nilsson L, Mohammed AK, Henriksson BG, Folkesson R, Winblad B, Bergstrom L. Environmental influence on somatostatin levels and gene expression in the rat brain. *Brain Research* 628(1–2): 93–98, 1993.

60. Hymovitch B. The effects of experimental variations on problem solving in the rat. *J. Comparative and Physiological Psychology* 45: 313–321, 1952.

61. Paylor R, Morrison SK, Rudy JW, Waltrip LT, Wehner JM. Brief exposure to an enriched environment improves performance on the Morris water task and increases hippocampal cytosolic protein kinase C activity in young rats. *Behavioural Brain Research* 52(1): 49–59, 1992.

62. Park GA, Pappas BA, Murtha SM, Ally A. Enriched environment primes forebrain choline acetyltransferase activity to respond to learning experience. *Neuroscience Letters* 143(1–2): 259–262, 1992.

63. Falkenberg T, Mohammed AK, Henriksson B, Persson H, Winblad B, Lindefors N. Increased expression of brain-derived neurotrophic factor mRNA in rat hippocampus is associated with improved spatial memory and enriched environment. *Neuroscience Letters* 138(1): 153–155, 1992.

64. Cornwell P, Overman W. Behavioral effects of early rearing conditions and neonatal lesions of the visual cortex in kittens. *J. Comparative and Physiological Psychology* 95(6): 848–862, 1981.

65. Krech D, Rosenzweig MR, Bennett EL. Relations between brain chemistry and problem-solving among rats raised in enriched and impoverished environments. *J. Comparative and Physiological Psychology* 55: 801–807, 1962.

66. Davenport RK Jr, Rogers CM, Menzel EW, Jr. Intellectual performance of differentially reared chimpanzees: II. Discrimination learning set. *American Journal of Mental Deficiency* 73: 963–969, 1967.

67. Escorihuela RM, Tobena A, Fernandez-Teruel A. Environmental enrichment reverses the detrimental action of early inconsistent stimulation and increases the beneficial effects of postnatal handling on shuttlebox learning in adult rats. *Behavioural Brain Research* 61(2): 169–173, 1994.

68. Davenport RK, Jr, Menzel EW, Jr. Stereotyped behavior of the infant chimpanzee. *Archives of General Psychiatry* 8: 99–104, 1963.

69. Davenport RK, Jr, Rogers CM. Intellectual performance of differentially reared chimpanzees. I. Delayed response. *American Journal of Mental Deficiency* 72: 674–680, 1968.

70. Stasiak M, Zernicki B. Delayed response learning to auditory stimuli is impaired in cage-reared cats. *Behavioural Brain Research* 53(1–2): 151–154, 1993.

71. Baddeley A. Working memory. *Science* 255: 556–559, 1992.

72. Just MA, Carpenter PA. A capacity theory of comprehension: Individual differences in working memory. *Psychological Review* 99: 122–149, 1992.

73. Squire LR. *Memory and Brain*. Oxford University Press, New York/Oxford, 1987.

74. Jensen AR. Spearman's g: Links between psychometrics and biology. *Annals of the New York Academy of Sciences* 702: 103–129, 1993.

75. Goldman-Rakic P. Circuitry of primate prefrontal cortex and regulation of behavior by representational memory. In *Handbook of Physiology: The Nervous System*. vol. V, ed. JR Pappenheimer. Krager, Bethesda, 1987, pp. 373–417.

76. Fuster JM, Alexander GE. Neuron activity related to short-term memory. *Science* 173: 652–654, 1971.

77. Kubota K, Niki H. Prefrontal cortical unit activity and delayed alternation performance in monkeys. *J. Neurophysiology* 34: 337–347, 1971.

78. Kojima S, Goldman-Rakic PS. Delay-related activity of prefrontal cortical neurons in rhesus monkeys performing delayed response. *Brain Research* 248: 43–49, 1982.

79. Menzel EW, Jr, Davenport RK, Jr, Rogers CM. The development of tool using in wild-born and restriction-reared chimpanzees. *Folia Primatologica* 12: 273–283, 1970.

80. Carpenter PA, Just MA, Reichle ED. Working memory and executive function: Evidence from neuroimaging. *Current Opinion in Neurobiology* 10: 195–199, 2000.

81. Rasmusson DD, Webster HH, Dykes RW. Neuronal response properties within subregions of raccoon somatosensory cortex 1 week after digit amputation. *Somatosensory and Motor Research* 9(4): 279–289, 1992.

82. Sharpless SK. Disuse supersensitivity. In *The Developmental Neuropsychology of Sensory Deprivation.* ed. AH Riesen. Academic Press, New York, 1975.

83. Eysel UT, Gonzalez-Aguilar F, Mayer U. Time-dependent decrease in the extent of visual deafferentiation in the lateral geniculate nucleus of adult cats with small retinal lesions. *Experimental Brain Research* 41(3–4): 256–263, 1981.

84. Gilbert CE, Wiesel TN. Receptive field dynamics in adult primary visual cortex. *Nature* 356(6365): 150–152, 1992.

85. Nicolelis MA, Lin RC, Woodward DJ, Chapin JK. Induction of immediate spatiotemporal changes in thalamic networks by peripheral block of ascending cutaneous information. *Nature* 361(6412): 533–536, 1993.

86. Kaas JH. Plasticity of sensory and motor maps in adult mammals. *Annual Review of Neuroscience* 14: 137–167, 1991.

87. Jenkins WM, Merzenich MM, Ochs MT, Allard T, Guic-Robles E. Functional reorganization of primary somatosensory cortex in adult owl monkeys after behaviorally controlled tactile stimulation. *J. Neurophysiology* 63(1): 82–104, 1990.

88. Clark SA, Allard T, Jenkins WM, Merzenich MM. Syndactyly results in the emergence of double-digit receptive fields in somatosensory cortex in adult owl monkeys. *Nature (London)* 332: 444–445, 1988.

89. Donoghue JP, Suner S, Sanes JN. Dynamic organization of primary motor cortex output to target muscles in adult rats. II. Rapid reorganization following motor nerve lesions. *Experimental Brain Research* 79(3): 492–503, 1990.

90. Sanes JN, Suner S, Lando JF, Donoghue JP. Rapid reorganization of adult rat motor cortex somatic representation patterns after motor nerve

injury. *Proceedings of the National Academy of Sciences of the United States of America* 85: 2003–2007, 1988.

91. Sanes JN, Suner S, Donoghue JP. Dynamic organization of primary motor cortex output to target muscles in adult rats. I. Long-term patterns of reorganization following motor or mixed peripheral nerve lesions. *Experimental Brain Research* 79(3): 479–491, 1990.

92. Merzenich M, personal communication, 2000.

93. Pons TP, Garraghty PE, Mishkin M. Lesion-induced plasticity in the second somatosensory cortex of adult macaques. *Proceedings of the National Academy of Sciences of the United States of America* 85: 5279–5281, 1988.

94. Kleinschmidt A, Beas MF, Singer W. Blockade of "NMDA" receptors disrupts experience-dependent plasticity of kitten striate cortex. *Science* 238: 355–358, 1987.

95. Rauschecker JP, Egert U, Kossel A. Effects of NMDA antagonists on developmental plasticity in kitten visual cortex. *International Journal of Developmental Neuroscience* 8: 425–435, 1990.

96. Daw NW. Mechanisms of plasticity in the visual cortex. The Friedenwald Lecture. *Investigative Ophthalmology and Visual Science* 35(15): 4168–4179, 1994.

97. Jiang CH, Tsien JZ, Schultz PG, Hu Y. The effects of aging on gene expression in the hypothalamus and cortex of mice. *Proceedings of the National Academy of Sciences of the United States of America* 98(4): 1930–1934, 2001.

98. Rampon C, Tsien JZ. Genetic analysis of learning behavior-induced structural plasticity [Review]. *Hippocampus* 10(5): 605–609, 2000.

99. Calossi A. Increase of ocular axial length in infantile traumatic cataract. *Optometry and Vision Science* 71(6): 386–391, 1994.

100. Bowering ER, Maurer D, Lewis TL, Brent HP. Sensitivity in the nasal and temporal hemifields in children treated for cataract. *Investigative Ophthalmalogy and Visual Science* 34(13): 3501–3509, 1993.

101. McCulloch DL, Skarf B. Pattern reversal visual-evoked potentials following early treatment of unilateral, congenital cataract. *Archives of Ophthalmology* 112: 510–518, 1994.

102. Birch EE, Swanson WH, Stager DR, Woody M, Everett M. Outcome after very early treatment of dense congenital unilateral cataract. *Investigative Ophthalmology and Visual Science* 34(13): 3687–3699, 1993.

103. Tytla ME, Lewis TL, Maurer D, Brent HP. Stereopsis after congenital cataract. *Investigative Ophthalmology and Visual Science* 34(5): 1767–1773, 1993.

104. von Senden M. *Space and Sight: The Perception of Space and Shape in the Congenitally Blind Before and After Operation* 1932. Reprint. Free Press, Glencoe, Ill., 1960.

105. Gregory RL, Wallace JG. Recovery from early blindness: A case study. *Quarterly Journal of Psychology.* 1963. Reprinted in *Concepts and Mechanisms of Perception.* ed. RL Gregory. Duckworth, London, 1974.

106. Sacks O. *Anthropologist on Mars. Seven Paradoxical Tales.* Knopf, New York, 1995, pp. 108–152.

107. Mishkin M, Appenzeller T. The anatomy of memory. *Scientific American* June 1987, pp. 80–89.

108. Rutstein RP, Fuhr PS. Efficacy and stability of amblyopin therapy. Optometry and vision. *Science* 69: 747–754, 1992.

109. Pascual-Leone A, Torres F. Plasticity of the sensorimotor cortex representation of the reading finger in Braille readers. *Brain* 116(Pt 1): 39–52, 1993.

110. Elbert T, Pentev C, Wienbruch C, Rockstroh B, Taub E. Increased cortical representation of the fingers of the left hand in string players. *Science* 270: 305–307, 1995.

111. Storfer MD. *Intelligence and Giftedness: The Contributions of Heredity and Early Environment.* Jossey-Bass, San Francisco, 1990.

112. Belmont L, Marolla FA. Birth order, family size, and intelligence. *Science* 182: 1096–1101, 1973.

113. Breland HM. Birth order, family configuration, and verbal achievement. *Child Development* 45: 1011–1019, 1974.

114. Jacobs BS, Moss HA. Birth order and sex of sibling as determinants of mother and infant interaction. *Child Development* 47: 315–322, 1976.

115. Lewis M, Kreitzberg VS. Effects of birth order and spacing on mother-infant interactions. *Developmental Psychology* 15: 617–625, 1979.

116. Gottfried AW, Gottfried AE. Home environment and cognitive development in young children of middle-socioeconomic families. In *Home Environment and Early Cognitive Development*. ed. AW Gottfried. Academic Press, Orlando, Fla., 1984.

117. White BL, Kaban BT, Attanucci J. *The Origins of Human Competence*. Lexington Books, Lexington, Mass., 1979.

118. Douglas JWB, Ross JM, Simpson HR. *All Our Future: A Longitudinal Study of Secondary Education*. Davies, London, 1968.

119. Page EB, Grandon GM. Family configuration and mental ability: Two theories contrasted with US data. *American Educational Research Journal* 16: 257–272, 1979.

120. Record RG, McKeown T, Edwards JH. The relation of measured intelligence to birth order and maternal age. *Annals of Human Genetics* 33: 61–69, 1969.

121. Zybert P, Stein Z, Belmont L. Maternal age and children's ability. *Perceptual and Motor Skills* 47: 815–818, 1978.

122. Skodak M, Skeels HM. A final follow-up study of one hundred adopted children. *Journal of Genetic Psychology* 75: 85–125, 1949.

123. Horn JM, Loehlin JC, Willerman L. Intellectual resemblance among adoptive and biological relatives: The Texas adoption project. *Behavior Genetics* 9: 177–207, 1979.

124. Schiff M. Intellectual status of working class children adopted early into upper middle-class families. *Science* 200: 1503–1504, 1978.

125. Flynn JR. Massive IQ gains in 14 nations: What IQ tests really measure. *Psychological Bulletin* 101(2): 171–191, 1987.

126. Horgan J. Get smart, take a test. *Scientific American* November 1995, pp. 12–14.

127. Schaie KW, Labourvie GV, Buech BO. Generational and cohort-specific differences in adult cognitive functioning: A fourteen-year study of independent samples. *Developmental Psychology* 9: 151–166, 1973.

128. Schaie KW. The Seattle longitudinal study: A 21-year exploration of psychometric intelligence in adulthood. In *Longitudinal Studies of Adult Psychological Development*. ed. KW Schaie. Guilford Press, New York, 1983.

129. Neisser U, Boodoo G, Bouchard TJ Jr, Boykin AW, Brody N, Ceci SJ, Halpern DF, Loehlin JC, Perloff R, Sternberg RJ, Urbina S. *Intelligence: Knowns and Unknowns. Report of a Task Force Established by the Board of Scientific Affairs of the American Psychological Association.* Science Directorate, Washington, D.C., 1995.

130. Terman LM, Merrill MA. *Measuring Intelligence: A Guide to Administration of the New Revised Stanford-Binet Tests of Intelligence.* Houghton Mifflin, Boston, 1937.

131. Seashore H, Wesman A, Doppelt J. Standardization of the Wechsler Intelligence Scale for Children. *Journal of Consulting Psychology* 14: 99–110, 1950.

132. Kaufman AS, Doppelt JE. Analysis of WISC-R standardization data in terms of the stratification variables. *Child Development* 47: 165–171, 1976.

133. Reynolds CR, Chastain RL, Kaufman AS, McLean JE. Demographic characteristics and IQ among adults: Analysis of the WAIS-R standardization sample as a function of the stratification variables. *Journal of School Psychology* 25: 323–342, 1987.

134. Ceci SJ. How much does schooling influence general intelligence and its cognitive components? A reassessment of the evidence. *Developmental Psychology* 27: 703–722, 1991.

135. Green RL, Hoffman LT, Morse R, Hayes ME, Morgan RF. *The Educational Status of Children in a District Without Public Schools (Cooperative Research Project No. 2321).* Office of Education, U.S. Department of Health, Education, and Welfare, Washington, D.C., 1964.

136. Teasdale TW, Owen DR. Thirty-year secular trends in the cognitive abilities of Danish male school-leavers at a high educational level. *Scandanavian Journal of Psychology* 35: 328–335, 1994.

137. Steinberg RJ. *Beyond IQ: A Triarchic Theory of Human Intelligence.* Cambridge University Press, New York, 1985.

138. Gould SJ. *The Mismeasure of Man.* W.W. Norton, New York, 1996.

139. Leiderman PH, Mendelson JH, Wexler D, Solomon P. Sensory deprivation: Clinical aspects. *Archives of Internal Medicine* 101: 389–396, 1958.

140. Lilly J. Mental effects of reduction of ordinary levels of physical stimuli on intact healthy persons. *Psychiatric Research Reports* 5: 1–9, 1956.

141. Shurley JT. Profound experimental sensory isolation. *American Journal of Psychiatry* 117: 539–545, 1960.

142. Zuckerman M, Albright RJ, Marks CS, Miller GL. Stress and hallucinatory effects of perceptual isolation and confinement. *Psychology Monographs* 76(30), 1962.

143. Zuckerman M, Persky H, Miller L, Levine B. Sensory deprivation versus sensory variation. *J. Abnormal Psychology* 76(1): 76–82, 1970.

144. Leff JP, Hirsch SR. The effects of sensory deprivation on verbal communication. *J. Psychiatric Research* 9: 329–336, 1972.

145. Jones A, Wilkinson H, Braden I. Information deprivation as a motivational variable. *J. Experimental Psychology* 62: 126–137, 1961.

146. Jones A. How to feed the stimulus hunger—problems in the definition of an incentive. Paper presented to the American Psychological Association, 1964.

147. Smith S, Myers TI. Stimulation seeking during sensory deprivation. *Perceptual and Motor Skills* 23: 1151–1163, 1966.

148. Berlyne DE. *Conflict, Arousal, and Curiosity*. McGraw-Hill, New York, 1960.

149. Bindra D. *Motivation: A Systematic Reinterpretation*. Ronald Press, New York, 1959.

150. Cofer CN, Appley MH. *Motivation: Theory and Research*. Wiley, New York, 1964.

151. Dember WN. *Psychology of Perception*. Holt, Rinehart, & Winston, New York, 1960.

152. Schultz DP. *Sensory Restrictions Effects on Behavior*. Academic Press, New York, 1965.

153. Harlow HF, Mears C. *The Human Model: Primate Perspectives*. V. H. Winston, Washington D.C., 1979.

154. Butler RA. The effect of deprivation of visual incentives on visual exploration in monkeys. *J. Comparative and Physiological Psychology* 50: 177–179, 1957.

155. Barnes GW, Kish GB, Wood WO. The effect of light intensity when onset or termination of illumination is used as reinforcing stimulus. *Psychological Record* 9: 53–60, 1959.

156. Barnes GW, Kish GB. Reinforcing properties of the onset of auditory stimulation. *J. Experimental Psychology* 62: 164–170, 1961.

157. Butler RA. Discrimination learning by rhesus monkeys to visual-exploration motivation. *J. Comparative and Physiological Psychology* 46: 95–98, 1953.

158. Fox S. Self-maintained sensory input and sensory deprivation in monkeys: A behavioral and neuropharmacological study. *J. Comparative Physiology and Psychology* 55: 438–444, 1962.

159. Schulman CA, Richlin M, Weinstein S. Hallucinations and disturbances of affect, cognition, and physical state as a function of sensory deprivation. *Perceptual and Motor Skills* 25: 1001–1024, 1967.

160. Heron W, Doane BK, Scott TH. Visual disturbances after prolonged perceptual isolation. *Canadian Journal of Psychology* 10: 13, 1956.

161. Zuckerman M, Hopkins TR. Hallucinations or dreams? A study of arousal levels and reported visual sensations during sensory deprivation. *Perceptual and Motor Skills* 22: 447–559, 1966.

162. Hayashi M, Morikawa T, Hori T. EEG alpha activity and hallucinatory experience during sensory deprivation. *Perceptual and Motor Skills* 75(2): 403–412, 1992.

163. Bexton WH, Heron W, Scott TH. Effects of decreased variation in the sensory environment. *Canadian Journal of Psychology* 8: 70, 1954.

164. Vernon J, McGill T. Sensory deprivation and pain thresholds. *Science* 133: 330–331, 1961.

165. Zubek JP, Flye J, Aftanas M. Cutaneous sensitivity after prolonged visual deprivation. *Science* 144: 1591–1593, 1964.

166. Zubek JP, Flye J, Willows D. Changes in cutaneous sensitivity after prolonged exposure to unpatterned light. *Psychonomic Science* 1: 283–284, 1964.

167. Smith S, Myers TI, Murphy DB. Vigilance during sensory deprivation. *Perceptual and Motor Skills* 24: 971–976, 1967.

168. Duda PD, Zubek JP. Auditory sensitivity after prolonged visual deprivation. *Psychonomic Science* 3: 359–360, 1965.

169. Zubek JP, Aftanas M, Hasek J, Sansom W, Schludermann E, Wilgosh L, Winocur G. Intellectual and perceptual changes during prolonged

perceptual deprivation: Low illumination and noise level. *Perceptual and Motor Skills* 15: 171–198, 1962.

170. Scott T, Bexton WH, Heron W, Doane BK. Cognitive effects of perceptual isolation. *Canadian Journal of Psychology* 13: 200–209, 1959.

171. Bruner JS. The cognitive consequences of early sensory deprivation. In *Sensory Deprivation*. eds. P. Solomon et al. Harvard University Press, Cambridge, Mass., 1961, pp. 195–207.

172. Zubek JP, Wilgosh L. Prolonged immobilization of the body: Changes in performance and in the electroencephalogram. *Science* 140: 306–308, 1963.

173. Zubek JP. Counteracting effects of physical exercises performed during prolonged perceptual deprivation. *Science* 142: 504–506, 1963.

174. Hebb DO. *The Organization of Behavior*. Wiley, New York, 1949.

Chapter 3

1. Harlow HF, Mears C. *The Human Model: Primate Perspectives*. V. II. Winston, Washington, D.C., 1979.

2. Rosenblatt JS, Siegel HI. In *Parental Care in Mammals*. eds. DJ Gubernick, PH Klopfer. Plenum, New York, 1981, pp. 1–76.

3. Insel TR, Young LJ. The neurobiology of attachment. *Nature Reviews Neuroscience* 2: 129–136, 2001.

4. Kleinman DG. Monogamy in mammals. *Quarterly Review of Biology* 52: 39–69, 1977.

5. Dewsbury DA. The comparative psychology of monogamy. In *American Zoology Nebraska Symposium on Motivation*. ed. DW Leger. University of Nebraska Press, Lincoln, 1988, pp. 1–50.

6. Ferguson JN, Young LJ, Insel TR. The neuroendocrine basis of social recognition. *Frontiers in Neuroendocrinology* 23: 200–224, 2002.

7. Breiter HC, Etcoff NL, Whalen PJ, Kenedy WA, Rauch SL, Buckner RL, Strauss MM, Hyman SE, Rosen BR. Response and habituation of the human amygdala during visual processing of facial expression. *Neuron* 17: 875–887, 1996.

8. Morris JS, Friston KJ, Buchel C, Frith CD, Young AW, Calder AJ, Dolan R. A neuromodulatory role for the human amygdala in processing emotional facial expressions. *Brain* 121: 47–57, 1998.

9. Schanberg SM, Field TM. Sensory deprivation stress and supplemental stimulation in the rat pup and preterm human neonate. *Child Development* 58: 1431–1447, 1987.

10. Spitz RA. Diacritical and coenesthetic organizations: The psychiatric significance of a functional division of the nervous system into a sensory and emotive part. *Psychoanalytic Review* 32: 146–161, 1945.

11. Blodgett FM. *Growth Retardation Related to Maternal Deprivation, Modern Perspectives in Child Development*. eds. A Solnit, S Provence. International Universities Press, Madison, Conn., 1963, pp. 83–97.

12. Powell GF, Brasel JA, Blizzard RM. Emotional deprivation and growth retardation. *New England Journal of Medicine* 276: 1271–1278; 1279–1283, 1967.

13. Field T, Schanberg SM, Scafidi F, Bauer CR, Vega-Lahr N, Garcia R, Nystrom J, Kuhn CM. Effects of tactile/kinesthetic stimulation on preterm neonates. *Pediatrics* 77: 654–658, 1986.

14. Scafidi F, Field T, Schanberg SM, Bauer C, Vega-Lahr N, Garcia R, Poirier J, Nystrom G, Kuhn CM. Effects of tactile/kinesthetic stimulation on the clinical course and sleep/wake behavior of preterm neonates. *Infant Behavior and Development* 9: 91–105, 1986.

15. Zhang LX, Levine S, Dent G, Zhan Y, Xing G, Okimoto D, Kathleen-Gordon M, Post RM, Smith MA. Maternal deprivation increases cell death in the infant rat. *Brain Research. Developmental Brain Research* 133: 1–11, 2002.

16. Lyons DM, Afariana H, Schatzberg AF, Sawyer-Glover A, Moseley ME. Experience-dependent asymmetric variation in primate prefrontal morphology. *Behavioral Brain Research* 136: 51–59, 2002.

17. Meaney MJ, Brake W, Gratton A. Environmental regulation of the development of mesolimbic dopamine systems: A neurobiological mechanism for vulnerability to drug abuse? *Psychoneuroendicrinology* 27: 127–138, 2002.

18. Siburg RM, Oitzl MS, Workel JO, de Kloet ER. Maternal deprivation increases 5-HT (1A) receptor expression in the CA1 and CA3 areas of senescent Brown Norway rats. *Brain Research* 912: 95–98, 2001.

19. Caldji C, Francis DD, Shasrma S, Plotsky PM, Meaney MJ. The effects of early rearing environment on the development of GABAA and central benzodiazepine receptor levels and novelty-induced fearfulness in the rat. *Neuropsychopharmacology* 22: 219–229, 2000.

20. Francis DD, Diorio J, Plotsky PM, Meaney MJ. Environmental enrichment reverses the effects of maternal separation on stress reactivity. *J. Neuroscience* 22: 7840–7843, 2002.

21. Kalinichev M, Easterling KW, Holtzman SG. Early neonatal experience of Long-Evans rats results in long-lasting changes in reactivity to a novel environment and morphine-induced sensitization and tolerance. *Neuropsychopharmacology* 27: 518–533, 2002.

22. Stephan M, Straub RH, Breivik T, Pabst R, von Horsten S. Postnatal maternal deprivation aggravates experimental autoimmune encephalomyelitis in adult Lewis rats: Reversal by chronic imipramine treatment. *International Journal of Developmental Neuroscience* 20: 125–132, 2002.

23. Weaver ICG, Cervoni N, Champagne FA, D'Alessio AC, Sharma S, Seckl JR, Szyf M, Meaney MJ. Epigenetic programming by maternal behavior. *Nature Neuroscience* 7: 847–854, 2004.

24. Weaver ICG, Diorio J, Seckl JR, Szyf M, Meaney MJ. Early environmental regulation of hippocampal glucocorticoid receptor gene expression: Characterization of intracellular mediators and potential genomic sites. *Annals of the New York Academy of Sciences* 1024: 182–212, 2004.

25. Fleming AS, Kraemer GW, Gonzalez A, Loveca V, Reesa S, Meloc A. Mothering begets mothering: The transmission of behavior and its neurobiology across generations. *Pharmacology, Biochemistry and Behavior* 73: 61–75, 2002.

26. Gonzalez A, Lovic V, Ward GR, Wainwright PE, Fleming AS. Intergenerational effects of complete maternal deprivation and replacement stimulation on maternal behavior and emotionality in female rats. *Developmental Psychobiology* 38: 11–32, 2001.

27. Jans JE, Woodside B. Effects of litter age, litter size and ambient temperature on the milk ejection reflex in lactating rats. *Developmental Psychobiology* 20: 333–344, 1987.

28. Kraemer GW. Psychobiology of early social attachment in rhesus monkeys. In *The Integrative Neurobiology of Affiliation*. eds. CS Carter, II Lederhendler, B Kirkpatrick, *Annals of the New York Academy of Sciences*, pp. 401–418, 1997.

29. Kraemer GW, Clarke AS. Social attachment, brain function and aggression. *Annals of the New York Academy of Sciences* 794: 121–135, 1996.

30. Ainsworth MDS. Attachment theory and its utility in cross-cultural research. In *Culture and Infancy. Variations in the Human Experience*. eds. PH Leiderman, SR Tulkin, A Rosenfeld. Academic Press, New York, 1977, pp. 49–68.

31. Carlson M, Earls F. Psychological and neuroendocrinological sequelae of early social deprivation in institutionalized children in Romania. In *The Integrative Neurobiology of Affiliation*. CS Carter, II Lederhendler, B Kirkpatrick, eds. *Annals of the New York Academy of Sciences*, pp 419–428, 1997.

32. Coplan JD, Trost RC, Owens MJ, Cooper TB, Gorman JM, Nemeroff CB, Rosenblum LA. Cerebrospinal fluid concentrations of somatostatin and biogenic amines in grown primates reared by mothers exposed to manipulated foraging conditions. *Archives of General Psychiatry* 55: 473–477, 1998.

33. Ferguson JN, Young LJ, Hearn EF, Matzuk MM, Insel TR, Winslow JT. Social amnesia in mice lacking the oxytocin gene. *Nature Genetics* 25: 284–288, 2000.

34. Isbell BJ, McKee L. Society's cradle: The socialization of cognition. In *Developmental Psychology and Society*, ed. J Saint. MacMillan, London, 1980, pp. 327–364.

35. Fifer WP, Moon CM. The role of mother's voice in the organization of brain function in the newborn. *Acta Paediatrica* 397(Suppl): 86–93, 1994.

36. Mehler J, Jusczyk P, Lambertz G, Halsted N, Bertoncini J, Amiel-Tison C. A precursor of language acquisition in young infants. *Cognition* 29: 143–178, 1988.

37. Mills M, Melhursh E. Recognition of mother's voice in early infancy. *Nature* 252: 123–124, 1974.

38. Goren CC, Sarty M, Wu PYK. Visual following and pattern discrimination of face-like stimuli by newborn infants. *Pediatrics* 56: 544–549, 1975.

39. Carpenter G. Mother's face and the newborn. *New Scientist* 21: 742–744, 1974.

40. Spitz R, Wolf K. The smiling response: A contribution to the ontogenesis of social relations. *Genetic Psychology Monographs* 34: 57–125, 1946.

41. Kaye K. *The Mental and Social Life of Babies. How Parents Create Persons.* University of Chicago Press, Chicago, 1982.

42. Tronick EZ (ed.) *Social Interchange in Infancy. Affect, Cognition, and Communication.* University Park Press, Baltimore, Md., 1982.

43. Sander L, Stechler G, Burns P, Julia H. Continuous 24-hour interactional monitoring in infants reared in two caretaking environments. *Psychosomatic Medicine* 34: 270–282, 1972.

44. Stern D. A microanalysis of mother–infant interaction: Behavior-regulating social contact between a mother and her 3½-month-old twins. *J. American Academy of Child Psychiatry* 10: 501–517, 1971.

45. Stern D. Mother and infant at play: The dyadic interaction involving facial, vocal, and gaze behaviors. In *The Effect of the Infant on Its Caregiver*. eds. M. Lewis, L. Rosenblum, Wiley, New York, 1974, pp. 187–214.

46. Fraiberg S. Blind infants and their mothers: An examination of the sign system. In *The Effect of the Infant on Its Caregiver*. eds. M. Lewis, L. Rosenblum. Wiley, New York, 1974, pp. 215–232.

47. Meltzhoff A, Moore MK. Imitation of facial and manual gestures by human neonates. *Science* 198: 75–78, 1977.

48. Emde R, Campos J, Reich J, Gaensbauer T. Infant smiling at five and nine months: Analysis of heart rate and movement. *Infant Behavior and Development* 1: 26–35, 1978.

49. Condon W, Sander L. Neonate movement is synchronized with adult speech: Interactional participation and language acquisition. *Science* 183: 99–101, 1974.

50. Strain B, Vietze P. Early dialogues: The structure of reciprocal infant-mother vocalization. Presented to the Society for Research in Child Development, March 1975.

51. Bruner JS. Learning the mother tongue. *Human Nature* 1: 42–49, 1978.

52. Yogman MW, Lester BM, Hoffman J. Behavioral and cardiac rhythmicity during mother–father–stranger–infant social interaction. *Pediatric Research* 17: 872–876, 1983.

53. Ter Vrugt D, Pederson DR. The effects of vertical rocking frequencies on the arousal level in two-month-old infants. *Child Development* 44: 205–209, 1973.

54. Korner AF, Grobstein R. Visual alertness as related to soothing in neonates: Implications for maternal stimulation and early deprivation. *Child Development* 37: 867–876, 1966.

55. Konner MJ. Aspects of the developmental ethology of a foraging people. In *Ethological Studies of Child Behavior*. ed. NB Jones. Cambridge University Press, London, 1972, pp. 285–304.

56. Brazeton TB. Joint regulation of neonate-parent behavior. In *Social Interchange in Infancy. Affect, Cognition, and Communication*. ed. EZ Tronick. University Park Press, Baltimore, Md., 1982, pp. 7–22.

57. Condon W, Sander L. Neonatal movement is synchronized with adult speech: Interactional participation and language acquisition. *Science* 183: 99–101, 1974.

58. Carter CS, Altemus M. Integrative functions of lactational hormones in social behavior and stress management. In *The Integrative Neurobiology of Affiliation*. eds. CS Carter, II Lederhendler, B Kirkpatrick. *Annals of the New York Academy of Sciences*, pp 164–174, 1997.

59. Keverne EB, Nevison CM, Martel FL. Early learning and the social bond. In *The Integrative Neurobiology of Affiliation*. eds. CS Carter, II Lederhendler, B Kirkpatrick. *Annals of the New York Academy of Sciences*, pp. 329–339, 1997.

60. Bibring CL, Dwyer TF, Huntington DS, Valenstein AF. A study of the psychological processes in pregnancy and of the earliest mother-child relationship. *Psychoanalytic Study of the Child* 16: 9–72, 1961.

61. White BL, Castle P, Held R. Observations on the development of visually guided directed reaching. *Child Development* 35: 349, 1964.

62. Kaye K. Organism, apprentice, and person. In *Social Interchange in Infancy. Affect, Cognition, and Communication*. ed. EZ Tronick. Baltimore, Md. University Park Press, 1982, pp. 183–194.

63. White B, Held R. Plasticity of sensorimotor development. In *Exceptional Infant*. vol 1. *The Normal Infant*. ed. J. Hellmut. Special Child Publications, Seattle, Wash., 1967.

64. Elbert T, Pentev C, Wienbruch C, Rockstroh B, Taub E. Increased cortical representation of the fingers of the left hand in string players. *Science* 270: 305–307, 1995.

65. Ceci SJ. How much does schooling influence general intelligence and its cognitive components? A reassessment of the evidence. *Developmental Psychology* 27: 703–722, 1991.

66. Seashore H, Wesman A, Doppelt J. Standardization of the Wechsler Intelligence Scale for children. *Journal of Consulting Psychology* 14: 99–110, 1950.

67. Eckerman C, Whatley J, McGehee LJ. Approaching and contacting the object another manipulates: A social skill of the 1-year-old. *Developmental Psychology* 15: 585–593, 1979.

68. Scaife M, Bruner JS. The capacity for joint visual attention in the infant. *Nature* 253: 265–266, 1975.

69. Morissette P, Ricard M, Gouin-Decarie T. Joint visual attention and pointing in infancy: A longitudinal study of comprehension. *British Journal of Developmental Psychology* 13: 163–176, 1995.

70. Corkum V, Moore C. The origins of joint visual attention in infants. *Developmental Psychology* 34: 28–38, 1998.

71. Murphy CM. Pointing in the context of a shared activity. *Child Development* 49: 371–380, 1978.

72. Bates E, Camaioni L, Volterra V. The acquisition of performatives prior to speech. *Merrill-Palmer Quarterly* 21: 205–226, 1975.

73. Butterworth G, Jarrett N. What minds have in common is space: Spatial mechanisms serving joint visual attention in infancy. *British Journal of Developmental Psychology* 9: 55–72, 1991.

74. Luria AR. *The Working Brain*. Trans. B Haugh. Basic Books, New York, 1973, pp. 265–268.

75. Hernandez-Péon R, Charver H, Jouvet H. Modification of electrical activity in the cochlear nuclei during attention in unanesthetized cats. *Science* 123: 331–332, 1956.

76. Woldroff MG, Liotti M, Seabolt M, Busse L, Lancaster JL, Fox PT. The temporal dynamics of the effects in occipital cortex of visual-spatial selective attention. *Cognitive Brain Research* 15: 1–15, 2002.

77. Vygotsky LS. *Mind in Society*, eds. M Cole, V John-Steiner, S Scribner, E Soubernam. Harvard University Press, Cambridge, Mass., 1978.

78. Friston KJ, Frith CD, Liddle PF, Fracowiak RS. Investigating a neural network model of word generation with positron emission tomography. *Proceedings of the Royal Society of London Ser. B* B244: 101–106, 1991.

79. Subrahmanyan K, Kraut RE, Greenfield PM, Gross EF. The impact of home computer use on children's activities and development. *Future of Children* 10: 123–144, 2000.

80. Merzenich MM, Jenkins WM, Johnston P, Schreiner C, Miller SL, Tallal P. Temporal processing deficits of language-learning impaired children ameliorated by training. *Science* 271(5245): 77–81, 1996.

81. Tallal P, Miller SL, Bedi G, Byma G, Wang X, Nagarajan SS, Schreiner C, Jenkins WM, Merzenich MM. Language comprehension in language-learning impaired children improved with acoustically modified speech. *Science* 271(5245): 81–84, 1996.

82. Wexler BE, Anderson M, Fulbright RK, Gore JC. Improved verbal working memory performance and normalization of task-related frontal lobe activation in schizophrenia following cognitive exercises. *American Journal of Psychiatry*, 157: 1094–1097, 2000.

83. Shweder RA. On savages and other children. *American Anthropologist* 84: 354–365, 1982.

84. Cole M. *Cultural Psychology.* Belknap Press of Harvard University Press, Cambridge, Mass., 1996.

85. Terrace H, Petitto L, Sanders R, Bever T. Can an ape create a sentence? *Science* 206: 891–899, 1979.

86. Williams JHG, Whiten A, Suddendorf T, Perrett DI. Imitation, mirror neurons and autism. *Neuroscience and Biobehavioral Reviews* 25: 287–295, 2001.

87. Rizzolatti G, Fadiga L, Gallese V, Fogassi L. Premotor cortex and the recognition of motor actions. *Cognitive Brain Research* 3(2): 131–141, 1996.

88. Umilta MA, Kohler E, Gallese V, Fogassi L, Fadiga L, Keysers C, Rizzolatti G. I know what you are doing. A neurophysiological study. *Neuron* 31(1): 1550–1565, 2001.

89. Iacoboni M, Woods RP, Brass M, Bekkering H, Mazziota JC, Rizzolatti G. Cortical mechanisms of human imitation. *Science* 286: 2526–2528, 1999.

90. Merians AS, Clark M, Poizner H, Macauley B, Gonzalez-Rothi LJ, Heilman K. Visual-imitative dissociation apraxia. *Neuropsychologia* 35: 1483–1490, 1997.

91. Goldenberg G, Hagman S. The meaning of meaningless gestures: A study of visuo-imitative apraxia. *Neuropsychologia* 35: 333–341, 1997.

92. Koski L, Wohlschlager A, Bekkering H, Woods RP, Dubeau MC, Mazziotta JC, Iacoboni M. Modulation of motor and premotor activity during imitation of target-directed actions. *Cerebral Cortex* 12: 847–855, 2002.

93. Meltzoff AN, Moore MK. Imitation of facial and manual gestures by human neonates. *Science* 198: 74–78, 1977.

94. Anderson B, Vietze P, Dokecki P. Reciprocity in vocal interactions of mothers and infants. *Child Development* 48: 1676–1681, 1977.

95. Stern D, Beebe B, Jaffe J, Bennett S. The infant's stimulus world during social interaction. In *Studies in Mother–Infant Interaction*. ed. HR Schaffer. Academic Press, London, 1977.

96. Klinnet M, Emde RN, Butterfield P, Campos JJ. Social referencing: The infant's use of emotional signals from a friendly adult with mother present. *Developmental Psychology* 22: 427–432, 1986.

97. Kuhl PK, Andruski JE, Chistovich IA, Chistovich LA, Kozhevnikova EV, Ryskina VL, Stolyarova EI, Sundberg U, Lacerda F. Cross-language analysis of phonetic units in language addressed to infants. *Science* 277: 684–686, 1997.

98. Kinsbourne M. The minor cerebral hemisphere as a source of aphasic speech. *Archives of Neurology* 25: 302–306, 1971.

99. Bishop DVM. Linguistic impairment after left hemidecortication for infantile hemiplegia? A reappraisal. *Quarterly J. Experimental Psychology* 35A: 199–208, 1983.

100. Zaidel E. Right hemisphere language. In *The Dual Brain*. eds. DF Benson, E Zaidel. Guilford Press, New York, 1985.

101. Dennis M. Capacity and strategy for syntactic comprehension after left or right hemidecortication. *Brain and Language* 10: 287–317, 1980.

102. Fenichel O. Identification. 1926. In *Pivotal Papers on Identification*. ed. GH Pollock. International Universities Press, Madison, Conn., 1993, pp. 57–74.

103. Freud S. Excerpt from Lecture XXXI: The dissection of the psychical personality. 1933. In *Pivotal Papers on Identification*. ed. GH Pollock. International Universities Press, Madison, Conn., 1993, pp. 47–52.

104. Greenson RR. The struggle against identification. 1954. In *Pivotal Papers on Identification*. ed. GH Pollock. International Universities Press, Madison, Conn., 1993, pp. 159–176.

105. Reich A. Early identifications as archaic elements in the superego. 1954. In *Pivotal Papers on Identification*. ed. GH Pollock. International Universities Press, Madison, Conn., 1993, pp. 177–196.

106. Ogden TH. *The Matrix of the Mind*. Jason Aronson, Northvale, N.J., 1986.

107. Winnicott DW. *Playing and Reality*. Basic Books, New York, 1971, p. 53.

108. Freud S. *An Outline of Psychoanalysis*. Trans. J Strachey. W.W. Norton, New York, 1949.

109. Hartmann H. *Ego Psychology and the Problem of Adaptation*. Trans. D Rapaport. International Universities Press, Madison, Conn., 1958.

110. Loewald H. *Papers on Psychoanalysis*. Yale University Press, New Haven, Conn., 1980.

111. Schafer R. Identification: A comprehensive and flexible definition. 1968. In *Pivotal Papers on Identification*. ed. GH Pollock. International Universities Press, Madison, Conn., 1993, pp. 305–346.

112. Freud S. The ego and the ID. In *The Standard Edition of the Complete Psychological Writings of Sigmund Freud*. Trans. J Strachey. vol. 19, pp. 38–39. Hogarth Press, London, 1966–74.

113. Erikson E. The problem of ego identity. 1959. In *Pivotal Papers on Identification*. ed. GH Pollock. International Universities Press, Madison, Conn., 1993, pp. 265–304.

114. Kernberg OF. Projection and projective identification: Developmental and clinical aspects. 1987. In *Pivotal Papers on Identification*. ed. GH Pollock. International Universities Press, Madison, Conn., 1993, pp. 405–426.

115. Müller-Schwarze D. Analysis of play behaviour: What do we measure and when? In *Play in Animals and Humans*. ed. PK Smith. Basil Blackwell, New York, 1984, pp. 147–158.

116. Parker ST. Playing for keeps: An evolutionary perspective on human games. In *Play in Animals and Humans*. ed. PK Smith. Basil Blackwell, New York, 1984, pp. 271–294.

117. Panskepp J, Beatty WW. Social deprivation and play in rats. *Behavioral and Neural Biology* 30: 197–206, 1980.

118. Bateson P, Martin P, Young M. Effects of interrupting cat mothers' lactation with bromocriptine on the subsequent play of their kittens. *Physiology and Behavior* 27: 845–845, 1981.

119. Martin P, Bateson P. The lactation-blocking drug bromocriptine and its application to studies of weaning and behavioral development. *Developmental Psychobiology* 15: 139–157, 1982.

120. Koepke JE, Pribram KH. Effect of milk on the maintenance of sucking behavior in kittens from birth to six months. *J. Comparative and Physiological Psychology* 75: 363–377, 1971.

121. Einon DF, Morgan MJ. A critical period for social isolation in the rat. *Developmental Psychobiology* 10: 123–132, 1977.

122. Rosenzweig MR. Effects of environment on the development of brain and behavior. In *Biopsychology of Development*. eds. E Tobach, LR Aronson, E Shaw. Academic Press, New York, 1971, pp. 303–342.

123. Morgan MJ. Effects of postweaning environment on learning in the rat. *Animal Behaviour* 21: 429–442, 1973.

124. Einon DF, Morgan MJ, Kibbler CC. Brief periods of socialization and later behavior in the rat. *Developmental Psychobiology* 11: 213–225, 1978.

125. Burghardt GM. *The Genesis of Animal Play: Testing the Limits*. MIT Press, Cambridge, 2005.

126. Fagen R. Play and behavioural flexibility. In *Play in Animals and Humans*. ed. PK Smith. Basil Blackwell, New York, 1984, pp. 159–174.

127. Millar S. Play. In *The Oxford Companion to Animal Behavior*. ed. DJ McFarland. Oxford University Press, New York, 1982, pp. 457–460.

Chapter 4

1. Hirt ER, Zillmann D, Erickson GA, Kennedy C. Costs and benefits of allegiance: Changes in fans' self-ascribed competencies after team victory versus defeat. *J. Personality and Social Psychology* 63: 724–738, 1992.

2. Cialdini RB, Borden RJ, Thorne A, Walker MR, Freeman S, Sloan LR. Basking in reflected glory: Three (football) field studies. *J. Personality and Social Psychology* 34: 366–375, 1976.

3. Cialdini RB, Richardson KD. Two indirect tactics of image management: Basking and blasting. *J. Personality and Social Psychology* 39: 406–415, 1980.

4. Bernhardt PC, Dabbs JM Jr., Fielden JA, Lutter CD. Testosterone changes during vicarious experiences of winning and losing among fans at sporting events. *Physiology and Behavior* 65(1): 59–62, 1998.

5. Hillman CH, Cuthbert BN, Cauraugh J, Schupp HT, Bradley MM, Lang PJ. Psychophysiological responses of sport fans. *Motivation and Emotion* 24(1): 13–28, 2000.

6. Gardner RW. The development of cognitive structures. In *Cognition: Theory, Research, Promise*. ed. C Scheere. Harper & Row, New York, 1964, pp. 147–171.

7. Kempler B, Wiener M. *Perception, Motives and Personality*. Knopf, New York, 1970.

8. Wolitzky DL, Wachtel PL. Personality and perception. In *Handbook of General Psychology*. ed. B Wolman. Prentice-Hall, Englewood Cliffs, N.J., 1973, pp. 826–855.

9. Gardner RW. Genetics and personality theory. In *Methods and Goals in Human Behavior Genetics*. ed. SG Vandenberg. Academic Press, New York, 1965, pp. 223–230.

10. Gardner RW, Moriarty A. *Personality Development at Preadolescence: Explorations of Structure Formations*. University of Washington Press, Seattle, 1968.

11. Nisbett RE, Peng K, Choi I, Norenzayan A. Culture and systems of thought: Holistic versus analytic cognition. *Psychological Review* 108(2): 291–310, 2001.

12. Nisbett RE. *The Geography of Thought*. Free Press, New York, 2003.

13. Spence DP, Holland B. The restricting effects of awareness: A paradox and an explanation. *J. Abnormal and Social Psychology* 64: 163–174, 1962.

14. Spence DP, Ehrenberg G. The effects of oral deprivation on responses to subliminal and supraliminal verbal food stimuli. *J. Abnormal and Social Psychology* 69: 10–18, 1964.

15. Gordon CM, Spence DP. The facilitating effects of food set and food deprivation on responses to a subliminal food stimulus. *J. Personality* 68: 409–416, 1966.

16. Postman L, Bruner J, McGinnies E. Personal values as selective factors in perception. *Psychological Review* 60: 298–306, 1953.

17. Bruner J, Goodman C. Value and need as organizing factors in perception. *J. Abnormal and Social Psychology* 42: 33–44, 1947.

18. Tajfel H. Value and the perceptual judgment of magnitude. *Psychological Review* 64: 192–204, 1957.

19. Kempler B, Wiener M. Personality and perception in the recognition threshold paradigm. *Psychological Review* 70: 349–356, 1963.

20. Postman L, Bruner J, McGinnies E. Personal values as selective factors in perception. *J. Abnormal and Social Psychology* 42: 143–154, 1948.

21. Fazio RH, Jackson JR, Dunton BC, Williams CJ. Variability in automatic activation as an unobtrusive measure of racial attitudes: A bona fide pipeline? *J. Personality and Social Psychology* 69: 1013–1027, 1995.

22. Whitenbrink B, Judd CM, Park R. Evidence for racial prejudice at the implicit level and its relationship to questionnaire measures. *J. Personality and Social Psychology* 72: 262–274, 1997.

23. Cunningham WA, Preacher KJ, Banaji MR. Implicit attitude measures: Consistency, stability and convergent validity. *Psychological Science* 121: 163–170, 2001.

24. Greenwald AG, McGhee DE, Schwartz JLK. Measuring individual differences in implicit cognition: The implicit association test. *J. Personality and Social Psychology* 74: 1464–1480, 1998.

25. Phelps EA, O'Connor KJ, Cunningham WA, Gatenby JC, Funayama ES, Gore JC, Banaji MR. Amygdala activation predicts performance on indirect measures of racial bias. *J. Cognitive Neuroscience* 12: 729–738, 2000.

26. Zajonc RB. Attitudinal effects of mere exposure. *J. Personality and Social Psychology Monograph* (Suppl.) 9(2): 1–27, 1968.

27. Winograd E, Goldstein FC, Monarach ES, Peluso JP, Goldman WP. The mere exposure effect in patients with Alzheimer's disease. *Neuropsychology* 13(1): 41–46, 1999.

28. Johnson MK, Kim JK, Risse G. Do alcoholic Korsakoff's syndrome patients acquire affective responses? *J. Experimental Psychology: Learning, Memory and Cognition* 11: 3–11, 1985.

29. Halpern AR, O'Connor MG. Implicit memory for music in Alzheimer's disease. *Neuropsychology* 14(3): 391–397, 2000.

30. Moreland RL, Zajonc RB. Exposure effects in person perception: Familiarity, similarity, and attraction. *J. Experimental Social Psychology* 18(5): 395–415, 1982.

31. Langlois JH, Roggman LA. Attractive faces are only average. *Psychological Science* 1: 115–121, 1990.

32. Gordon PC, Holyoak KJ. Implicit learning and generalization of the "mere exposure" effect. *J. Personality and Social Psychology* 45: 492–500, 1983.

33. Kunst-Wilson WR, Zajonc RB. Affective discrimination of stimuli that cannot be recognized. *Science* 207(4430): 557–558, 1980.

34. Bornstein RF, D'Agostino PR. Stimulus recognition and the mere exposure effect. *J. Personality and Social Psychology* 63(4): 545–552, 1992.

35. Seamon JG, Ganor-Stern D, Crowley MJ, Wilson SM, Weber WJ, O'Rourke CM, Mahoney JK. A mere exposure effect for transformed three-dimensional objects: Effects of reflection, size, or color changes on affect and recognition. *Memory and Cognition* 25(3): 367–374, 1997.

36. Peretz I, Gaudreau D, Bonnel AM. Exposure effects on music preference and recognition. *Memory and Cognition* 26(5): 884–902, 1998.

37. Willems S, Adam S, Van der Linden M. Normal mere exposure effect with impaired recognition in Alzheimer's disease. *Cortex* 38(1): 77–86, 2002.

38. Alluisi EA, Adams OS. Predicting letter preferences: Aesthetics and filtering in man. *Perceptual and Motor Skills* 14: 124–131, 1962.

39. Hoorens V, Nuttin JM. Overvaluation of own attributes: Mere ownership or subjective frequency? *Social Cognition* 11(2): 177–200, 1993.

40. Zajonc R. Brainwash: Familiarity breeds comfort. *Psychology Today* 3(9): 32–35, 60–64, 1970.

41. Saegert S, Swap W, Zajonc RB. Exposure, context, and interpersonal attraction. *J. Personality and Social Psychology* 25(2): 234–242, 1973.

42. Festinger L. *Conflict, Decision, and Dissonance.* Stanford University Press, Stanford, Calif., 1964.

43. Festinger L. *A Theory of Cognitive Dissonance.* Stanford University Press, Stanford, Calif., 1957.

44. Hastorf A, Cantril H. They saw a game: A case study. *J. Abnormal and Social Psychology* 49: 129–134, 1954.

45. Croyle RT, Cooper J. Dissonance arousal: Physiological evidence. *J. Personality and Social Psychology* 45(4): 782–791, 1983.

46. Elkin RA, Leippe MR. Physiological arousal, dissonance, and attitude change: Evidence for a dissonance-arousal link and a "don't remind me" effect. *J. Personality and Social Psychology* 51(1): 55–65, 1986.

47. Lewin K. Group decision and social change. In *Readings in Social Psychology.* eds. G Swanson, T Newcomb, E Hartley. Henry Holt, New York, 1952.

48. Lieberman S. The effects of changes in roles on the attitudes of role occupants. In *Human Behavior and International Politics.* ed. JD Singer. Rand McNally, Chicago, 1965.

49. Schachter S, Burdick H. A field experiment on rumor transmission and distortion. *J. Abnormal and Social Psychology* 50: 363–371, 1955.

50. Asch SE. Studies of independence and conformity: I. A minority of one against a unanimous majority. *American Psychological Association,* Washington, 1956.

51. Milgram S. *The Individual in a Social World: Essays and Experiments.* Addison-Wesley, Reading, Mass., 1977.

52. Tronick E, Als H, Adamson L, Wise S, Brazelton TB. The infant's response to entrapment between contradictory messages in face-to-face interaction. *J. American Academy of Child Psychiatry* 17: 1–13, 1978.

53. Cohn JF, Tronick EZ. Communicative rules and sequential structure of infant behavior during normal and depressed interaction. In *Social Interchange in Infancy*. ed. EZ Tronick. University Park Press, Baltimore, Md., 1982.

54. Tronick EZ, Als H, Adamson L. Structure of early face-to-face communicative interaction. In *Before Speech: The Beginnings of Human Comunication*. ed. M Bullowa. Cambridge University Press, Cambridge, 1979.

55. Harmon DK, Masuda M, Holmes TH. The social readjustment rating scale: A cross-cultural study of Western Europeans and Americans. *J. Psychosomatic Research* 14: 391–400, 1970.

56. Jacobs S, Douglas L. Grief: A mediating process between a loss and illness. *Comprehensive Psychiatry* 20: 165–176, 1979.

57. Parkes CM. *Bereavement: Studies of Grief in Adult Life*. International Universities Press, Madison, Conn., 1972.

58. Bowlby J. *Attachment and Loss* vol. 3, *Separation, Anxiety, and Anger*. Hogarth Press, London, 1973.

59. Prigerson HG, Maciejewski PK, Newsom J, Reynolds CF, Frank E, Bierhals AJ, Miller MD, Fasiczka A, Doman J, Houck PR. The Inventory of Complicated Grief: A scale to measure maladaptive symptoms of loss. *Psychiatry Research* 59: 65–79, 1995.

60. Freud S. Mourning and melancholia. In *The Standard Edition of the Complete Psychological Writings of Sigmund Freud*. Trans. J Strachey. vol. 14, pp. 243–258, Hogarth Press, London, 1966–74.

61. Jacobs S, Ostfeld A. An epidemiological review of the mortality of bereavement. *Psychosomatic Medicine* 39(5): 344–357, 1977.

62. Rees WD, Lutkin SG. Mortality of bereavement. *British Medical Journal* 4: 13–16, 1940.

63. Prigerson HG, Bierhals AJ, Kasl SV, Reynolds CF, Shear MK, Day N, Beery LC, Newsom JT, Jacobs S. Traumatic grief as a risk factor for mental

and physical morbidity. *American Journal of Psychiatry* 154(5): 617–612, 1997.

64. Hoffman E. *Lost in Translation*. Penguin Books, New York, 1989.

65. Kincaid J. "Poor Visitor" in *Lucy*. Farrar Strauss, New York, 2002.

66. Krystal H, Petty TA. Dynamics of adjustment to migration. *Proceedings of the Third World Congress of Psychiatry, Psychiatric Quarterly* (Suppl.) 37: 118–133, 1963.

67. Garza-Guerrero AC. Culture shock: Its mourning and the vicissitudes of identity. *J. American Psychoanalytic Association* 22: 408–429, 1974.

68. Amati-Mehler J, Artentieri S, Canestri J. *The Babel of the Unconscious: Mother Tongue and Foreign Languages in the Psychoanalytic Dimension*. Trans. J Whitelaw-Cucco. International Universities Press, Madison, Conn., 1993.

69. Volkan VD. Immigrants and refugees: A psychodynamic perspective. *Mind and Human Interaction* 4: 63–69, 1993.

70. Akhtar S. A third individuation: Immigration, identity, and the psychoanalytic process. *J. American Psychoanalytic Association* 43(4): 1051–1085, 1995.

Chapter 5

1. Whiten A, Goodall J, McGrew WC, Nishida T, Reynolds V, Sugiyama Y, Tutin CEG, Wrangham RW, Boesch C. Cultures in chimpanzees. *Nature* 399: 682–685, 1999.

2. Hirata S, Morimura N. Naïve chimpanzees (*Pan troglodytes*) observation of experienced conspecifics in a tool-using task. *J. Comparative Psychology* 114: 291–296, 2000.

3. Goodall J. *Reason for Hope: A Spiritual Journey*. Warner Books, New York, 1999.

4. Robertson I. *Sociology*. Worth Publishers, New York, 1987.

5. Myers DG. *Social Psychology.* 6th ed. McGraw-Hill, New York, 1999.

6. McGranahan DV, Wayne I. German and American traits reflected in popular drama in human behavior and international politics: contributions from the social-psychological sciences. In *Human Behavior and International Politics; Contributions from the Social-Psychological Sciences.* ed. JD Singer. Rand McNally, Chicago, 1965, pp. 123–135.

7. Stoodley BH. Normative attitudes of Filipino youth compared with German and American youth. *American Sociological Review* 22: 553–561, 1957.

8. Triandis HC. *Culture and Social Behavior,* McGraw-Hill, New York, 1994.

9. Vogt EZ, O'Dea TF. A comparative study of the role of values in social action in two southwestern communities. *American Sociological Review* 18: 645–654, 1953.

10. Christensen HT. Cultural relativism and premarital sex norms. *American Sociological Review* 25: 31–39, 1960.

11. Sontag D. Defiant muslims begin building Nazareth mosque. *New York Times,* International section, November 24, 1999, p. 3.

12. Hart AJ, Whalen PJ, Shin LM, McInerney SC, Fischer H, Rauch SL. Differential response in the human amygdala to racial outgroup vs. ingroup face stimuli. *NeuroReport* 11: 2351–2355, 2000.

13. Pratt ML. *Imperial Eyes: Travel Writing and Transculturation.* Routledge, London, 1992.

14. Jacobsen M. *Barbarian Virtues: The United States Encounters Foreign People at Home and Abroad.* Hill and Wang, New York, 2000.

15. Bradford PV, Blume H. *Ota Benga: The Pygmy in the Zoo.* Dell Publishing, New York, 1992.

16. Wexler L. *Tender Violence.* University of North Carolina Press, Chapel Hill, 2000.

17. *Encyclopedia Americana*, 1903 edition.

18. Cole, M. *Cultural Psychology*. Belknap Press of Harvard University Press, Cambridge, Mass., 1996.

19. James H. *The American Scene*. Indiana University Press, Bloomington, 1968.

20. Gourevitch P. *We wish to inform you that tomorrow we will be killed with our families: Stories from Rwanda*. Farrar Straus and Giroux, New York, 1998.

21. Sahlins M. *How "Natives" Think: About Captain Cook, for Example*. University of Chicago Press, Chicago, 1995.

22. Diamond J. *Collapse: How Societies Choose to Fail or Succeed.* Viking, New York, 2005.

23. Cronon W. *Changes in the Land: Indians, Colonists, and the Ecology of New England*. Hill and Wang, New York, 1983.

24. Herbert Z. *Barbarian in the Garden*. Trans. M March, J Anders. Carcanet, Manchester, UK, 1985.

25. Lesourd P, Ramiz JM. *On the Path of the Crusaders*. Massada Press, Jerusalem, 1969.

26. Runciman S. *The First Crusade*. Cambridge University Press, New York, 1980.

27. Finucane RC. *Soldiers of the Faith: Crusaders and Moslems at War*. J.M. Dent & Sons, London, 1983.

28. Riley-Smith JSC. *First Crusade and the Idea of Crusading*. Athline Press, London, 1986.

29. Hallam E. (ed.) *Chronicles of the Crusades: Eye-Witness Accounts of the Wars between Christianity and Islam*. Weidenfeld and Nicolson, London, 1989.

30. Riley-Smith L, Riley-Smith JSC. *The Crusades Idea and Reality, 1095– 1274*. Edward Arnold, London, 1981.

31. Tyerman C. *Fighting for Christendom Holy War and the Crusades.* Oxford University Press, Oxford, 2004.

32. Phillips J. *The Fourth Crusade and the Sack of Constantinople.* Viking, New York, 2004.

33. Asbridge T. *The First Crusade A New History.* Oxford University Press, Oxford, 2004.

34. Goody JR. *Islam in Europe.* Polity Press, Cambridge, UK, 2004.

35. Kaplan RD. *Balkan Ghosts a Journey through History.* St. Martin's Press, New York, 1993.

36. Sells MA. *The Bridge Betrayed: Religion and Genocide in Bosnia.* University of California Press, Berkeley, 1996.

Epilogue

1. Wurm SA (ed.) *Atlas of the World's Languages in Danger of Disappearing.* UNESCO, Paris, 1996.

2. Davis W. Vanishing cultures. *National Geographic* August: 62–89, 1999.

3. United Nations press release GA/SHC/3488, October 28, 1998.

4. Goodenough OR. Defending the imaginary to the death? Free trade, national identity and Canada's cultural preoccupation. *Arizona Journal of International and Comparative Law*, pp. 203–253, Winter 1998.

5. Kaplan LGC. The European Community's television without fantasies directive: Stimulating Europe to regulate curfew. *Emory International Law Review*, pp. 255–346, Spring 1994.

6. Kim CH. Building the Korean film industry's competitiveness: Abolish the screen quota and subsidize the film industry. *Pacific Rim Law and Policy Journal*, pp. 353–378, May 2000.

7. Cohen R. Fearful over the future, Europe seizes on food. *New York Times*, section 4A, page 1, August 29, 1999.

8. Vanstan, C. In search of the mot juste: Re Touban law and the European Union. *Boston College International and Comparative Law Review*, pp. 175–194, Winter 1999.

9. Daley S. Use of English as world language is booming, and so is concern. *New York Times*, front section, page 1, April 16, 2001.

10. Kiernen B. *Pol Pot Regime: Race, Power and Genocide in Cambodia under the Khmer Rouge 1975–79*. Yale University Press, New Haven, Conn., 1996.

11. Watkin H. Hanoi intensifies censorship laws in the name of culture. *South China Morning Post*, August 1, 2000.

12. Ambach FS. Arabs riding Harleys rev up Emirates ire. *Christian Science Monitor* 87:99, April 18, 1995.

13. Geertz C. *The Interpretation of Cultures*. Basic Books, New York, 1973.

14. Numan M, Sheehan TP. Neuroanatomal circuitry for mammalian maternal behavior. *Annals of the New York Academy of Sciences* 807: 101–225, 1997.

15. Numan M. A neural circuitry analysis of maternal behavior in the rat. *Acta Paediatrica* (Suppl.) 397: 19–28, 1994.

16. Modney BK, Hatton GI. Maternal behaviors: Evidence that they feed back to alter brain morphology and function. *Acta Paediatrica* (Suppl.) 397: 29–32, 1994.

17. Insel TR, Young LJ. The neurobiology of attachment. *Nature Reviews Neuroscience* 2: 129–136, 2001.

18. Bibring CL, Dwyer TF, Huntington DS, Valenstein AF. A study of the psychological processes in pregnancy and of the earliest mother–child relationship. *Psychoanalytic Study of the Child* 16: 9–72, 1961.

19. Wingert P, Laverman JF. Parents behaving badly. *Newsweek* 136: 47, July 14, 2000.

20. National Association of Sports Officials website (www.naso.org).

21. Demos J. Presentation to Genocide Studies Program. Yale University, 2001.

22. Gaspard G. *A Small City in France*. Trans. A Goldhammer. Harvard University Press, Cambridge, Mass., 1995.

23. Buruma I. Final cut. *New Yorker*, pp. 26–32, January 3, 2005.

24. Snow CP. *The Two Cultures and the Scientific Revolution*. Cambridge University Press, Cambridge, 1959.

Index